CLIMATE ACTION
Mark Diesendorf

DR MARK DIESENDORF is deputy director of the Institute of Environmental Studies at University of New South Wales, Sydney. Previously he was a principal research scientist in CSIRO, Australia's national scientific research organisation, and professor of Environmental Science at the University of Technology, Sydney. His work in community and professional groups included being president of the original Australasian Wind Energy Association, coordinator of the Australian Conservation Foundation's Climate Change Program, and vice-president of the Australia New Zealand Society for Ecological Economics. He was also an active member of several NGOs devoted to environment, health, renewable energy, appropriate technology and peace. He is co-editor of *Human Ecology, Human Economy: Ideas for an ecologically sustainable future* and author of *Greenhouse Solutions with Sustainable Energy*.

Disclaimer

This book is intended to provide basic information for individuals, groups and organisations on various nonviolent strategic and tactical options that may be available for pressuring governments and businesses to take action to abate climate change. The publisher, author, editors and endorsers of this publication do not encourage or condone tactics that involve the use of any kind of force or violence.

In some jurisdictions, the tactics described here may not be lawful. Potential legal risks may include civil law suits (eg, being sued for defamation, trespass, nuisance, interference with business activities), minor penalties (eg, fines, cautions), or even criminal prosecutions (eg, for offences such as obstruction, unlawful assembly). The publisher, author, editors and endorsers of this publication do not condone or encourage the carrying out of any activities that are unlawful. It is the responsibility of the reader to consider the legality of any proposed tactic and its potential consequences, and to seek legal advice before taking any action. The publisher, author, editors and endorsers expressly disclaim all and any liability and responsibility to any person for anything done in reliance, whether wholly or partially, on the whole or any part of the contents of this publication.

Note: Terms defined in the Glossary are italicised where they first appear in the text. The Glossary can be found at the back of the book.

Climate Action

A campaign manual for greenhouse solutions

Mark Diesendorf

Illustrated by Simon Kneebone

UNSW
PRESS

A UNSW Press book

Published by
University of New South Wales Press Ltd
University of New South Wales
Sydney NSW 2052
AUSTRALIA
www.unswpress.com.au

Text © Mark Diesendorf 2009
Cartoons © Simon Kneebone 2009
First published 2009

National Library of Australia
Cataloguing-in-Publication entry

    Author: Mark Diesendorf
    Title: Climate action: a campaign manual for greenhouse solutions/
    Mark Diesendorf.
    ISBN: 978 1 74223 018 4 (pbk.)
    Notes: Includes index.
    Subjects: Social action – Australia – Handbooks, manuals, etc.
        Pressure groups – Australia – Handbooks, manuals, etc.
        Political participation – Australia – Handbooks, manuals, etc.
        Environmentalists – Australia – Political activity – Handbooks, manuals, etc.
        Citizens' associations – Australia – Handbooks, manuals, etc.
        Campaign management – Australia – Handbooks, manuals, etc.
        Climatic changes – Social aspects
        Greenhouse gas mitigation
    Dewey Number: 322.43068

Design Josephine Pajor-Markus
Cover Design by Committee
Printer Ligare

FSC
Mixed Sources
Product group from well-managed
forests and other controlled sources
Cert no. SGS-COC-004233
www.fsc.org
© 1996 Forest Stewardship Council

The paper this book is printed on is certified by the
©1996 Forest Stewardship Council (FSC). The FSC
promotes environmentally responsible, socially beneficial
and economically viable management of the world's forests.

CONTENTS

Contents

PREFACE

This book is for people who are concerned about global climate change and want to do something about it. It is for people who care for the future of humanity and more specifically for the lands and their inhabitants that will suffer, or are already suffering, from floods, storm surges, droughts, heatwaves, wildfires and loss of biodiversity resulting from the industries and other human activities that emit greenhouse gases. It is for people who are concerned for their own communities and for the children and grandchildren of the present generation.

As a first step in reducing our greenhouse gas emissions and those of our families, we can purchase and use appropriate products and modify our lifestyles. We may, for example, be using an energy efficient refrigerator, water efficient showerheads and taps, an insulating blanket over the hot water tank, fluorescent lights, and an electric fan or evaporative cooler instead of air conditioning. We may be purchasing *Green Power* from an accredited *energy* retailer, taking fewer trips by car and more trips by walking and cycling, eating less beef and lamb, and wearing jumpers in winter instead of heating our homes excessively. Those of us who are fortunate to own our own homes may be planning to insulate them properly, to install a hot water system based on solar, *gas* or electric heat pump, and possibly even a solar electricity system.

All such household actions are helpful, but they are not nearly enough to avert the climate crisis. It is governments and big business that make the key decisions that determine greenhouse gas emissions: to grant permission for either a new coal-fired power station or wind and solar farms; to build a new motorway or a railway; to integrate urban and transport planning or 'to leave it to the market', that is, to the developers. It is governments that collect our taxes and decide how to spend them and it is the big greenhouse gas emitting industries that exert strong influence over government decisions.

This book is for those of you who wish to go beyond individual and household actions, to help to transform community awareness of the climate crisis and to exercise your democratic rights to put your concerns peacefully but firmly to government and business between elections. You will be most effective if you join a group that is part of the *climate action movement* and give some of your time, knowledge, skills and possibly money to the cause. A wide variety of groups belong to the growing climate action movement: large environmental and social justice *non-government organisations* (NGOs); professional, business, trade union, student and faith-based groups; and small local community groups dedicated entirely to climate action. Becoming active in one of these groups can give you new skills, new friends, mutual support, and a sense of progress and achievement. Even if you prefer not to join a group, perhaps because you live somewhere remote or have family constraints, you can still use the ideas and experience this book brings together.

I began my career as a physical scientist and later broadened out to become an interdisciplinary environmental studies educator and researcher. I came to understand that scholarship is necessary but not sufficient for solving the major problems facing human society. We also need community action to counter the ignorance, greed and lust for power that drives the destructive tendencies in humans. I'm inspired by Mahatma Gandhi, Martin Luther King Jr, David Suzuki, Vandana Shiva, James Hansen and the individuals and groups that have won the Right Livelihood Award, sometimes referred to as the Alternative Nobel Prize. So, over several decades, I have been active in community groups concerned with social responsibility in science, environmental protection, health, appropriate technology, renewable energy, green business and peace.

This book starts by explaining briefly in non-specialist language why climate change is serious and urgent – why it can truly be called a crisis or emergency. Next I shatter myths being disseminated by vested interests, as they attempt to undermine climate action. Then I show that we already have most of the technologies needed to cut green-house gas emissions dramatically. But most governments have not enacted effective policies to implement these technologies on a large scale, preferring instead rhetoric and token projects. The essential policies for stopping global warming are explained. There is no single 'magic bullet', but rather a portfolio of vital actions. Then we come to the core of the book: the strategies and *tactics* available to groups and individuals to inform the community and exert pressure on the power-holders to change direction towards an ecologically sustainable and socially just future.

ACKNOWLEDGMENTS

I thank Regina Betz, Dave Elliott, Nina Hall, Iain MacGill, Brian Martin, Barrie Pittock, Sue Wareham and the anonymous reviewers for valuable comments on parts of the manuscript and/or answering my persistent questions. Brian Martin in particular devoted much thought to the whole manuscript. Copy-editor Jessica Perini made many constructive suggestions for improving the content and presentation. I'm grateful to Gabriella Sterio from UNSW Press for encouraging me to write this book. The quality of the book has been greatly enhanced by the advice of all these readers. However, the opinions expressed in the book, errors and omissions are my own responsibility.

I'm grateful to Nina Hall for writing the appendices; Philippa Rowland for box 6.1; Jim Green from Friends of the Earth for permission to reproduce his media release in box 6.2; Regina Betz, Iain MacGill and the Centre for Policy Development for permission to reproduce the material that appears in box 4.1; the Midwest Academy for permission to use its Strategy Chart as the template for table 5.1; New Society Publishers for permission to publish the material in table 5.2; and the Environmental Defenders Office (NSW) for legal advice.

Many thanks to Simon Kneebone for his powerful cartoons which capture the spirit of this book. Simon has been a freelance cartoonist and illustrator for over 20 years; his cartoons have appeared in many publications in the areas of social issues, the environment, agriculture and education.

I dedicate this book to my sons, Thor, Danny and Joey, and grandchildren Bede and Asha.

UNITS AND CONVERSION FACTORS

Powers of 10

Prefix	Symbol	Value	Example
kilo	k	10^3	kilowatt kW
mega	M	10^6	megawatt MW
giga	G	10^9	gigajoule GJ
tera	T	10^{12}	terawatt-hour TWh
peta	P	10^{15}	petajoule PJ

SI units

Basic unit	Name	Symbol
length	metre	m
mass	kilogram	kg
time	second	s
temperature	degree Kelvin	°K

Derived unit	Name	Symbol
energy	joule	J
power	watt	W
potential difference	volt	V
temperature	degree Celsius	°C
time	hour	h

Conversion factors

Type	Name	Symbol	Value
volume	cubic metre	m^3	1000 litres = 1 kL
volume	gallon (US)		3.785 L
energy	kilowatt-hour	kWh	3.6×10^6 J = 3.6 MJ
energy	terawatt-hour	TWh	3.6×10^{15} J = 3.6 PJ
energy	litre of petrol	L	3.2×10^7 J
energy	barrel of oil		6.12×10^9 J = 6.12 GJ
energy	barrel of oil		159 L
energy	British thermal unit	btu	1055.1 J = 1.0551 kJ
energy	cubic metre of natural gas at STP*	m^3	3.4×10^7 J
energy	tonne of black coal	t	23 GJ
energy	tonne of brown coal	t	10 GJ
energy	tonne of green wood	t	10 GJ
energy	tonne of oven-dried wood	t	20 GJ
carbon	tonne of carbon	t	3.67 t CO_2
power	kWh per year	kWh/y	0.114 W
time	year	y	8760 hours
temperature	degree Celsius	°C	(°C + 273) °K
speed	kilometre per hour	km/h	0.2778 m/s

*STP is 'standard temperature & pressure', defined as 1 atmosphere of pressure and 0 degrees Celsius.

ABBREVIATIONS

A$	Australian dollars
c/kWh	cents per kilowatt-hour
CCS	carbon (dioxide) capture and sequestration or storage
CDM	Clean Development Mechanism
CMO	climate movement organisation
CO_2	carbon dioxide
COAG	Council of Australian Governments
CSIRO	Commonwealth Scientific and Industrial Research Organisation (Australia)
EITE	emissions-intensive trade-exposed
EU	European Union
FIT	feed-in tariff
GDP	gross domestic product
HCFC	hydrochlorofluorocarbon
IPCC	Intergovernmental Panel on Climate Change
MAP	Movement Action Plan
MRET	Mandatory Renewable Energy Target
NGO	non-government organisation
OECD	Organisation for Economic Cooperation and Development
OOA	Organization for Information on Nuclear Power (Denmark)
PFC	perfluorocarbon
PV	photovoltaic
REDD	reduce deforestation and degradation in tropical rainforests
RGGI	Regional Greenhouse Gas Initiative
RPS	renewable (energy) portfolio standard
UNFCCC	United Nations Framework Convention on Climate Change
US$	US dollars
WTO	World Trade Organisation

1

THREAT AND HOPE

Deadly firestorms in Victoria and California; an extraordinary heatwave killing tens of thousands in Europe; torrential floods in the United Kingdom and Queensland; more frequent severe hurricanes or cyclones; long droughts; plants flowering earlier and disrupting the life-cycles of insects and animals; warming of the Earth's surface and lower atmosphere, coupled with cooling of the stratosphere; vast ice shelves collapsing into the polar oceans. Taken singly, most of these phenomena could be the result of natural fluctuations in weather. Taken together, they form and identify the fingerprint of a criminal sabotaging the world's climate systems. That criminal is us human beings, or, more precisely, some of us.

Global warming is one of the greatest threats to civilisation that humans have ever known. But this should not be a cause for paralysis or panic. The threat has solutions, most of which are available now. Under the influence of powerful vested interests that are opposing change, most governments and businesses are not implementing these solutions beyond a token level. For those of us who are concerned, there is a way forward. It involves growing and empowering the community-based climate action movement until no government or business can resist it. This book aims to assist you and your *climate action group* in that process.

The threat

As an environmental scientist and a citizen, I'm concerned that we humans are destroying our planet in front of our own eyes. We are just like the inhabitants of Easter Island who, for petty competitive reasons, cut down the last trees on the island and so ended the building of fishing boats and the very basis for their economy and society.[1] The scientific evidence for human-induced climate change is compelling. The Intergovernmental Panel on Climate Change states that 'most of the observed increase in globally averaged temperatures since the mid-20th century is *very likely* due to the observed increase in *anthropogenic* greenhouse gas concentrations'.[2]

Global climate change, caused primarily by the emission of greenhouse gases from human activities, is accelerating. Temperature increases, averaged over 10-year periods, have accelerated through the late 20th century. Ten of the 12 years 1997–2008 inclusive have been the warmest recorded by instruments.[3] Melting of the Arctic ice cap and the vast majority of this planet's glaciers and snowfields is also accelerating. In 2006 climate scientists generally believed that the Arctic ice cap might have totally melted by the Summer of 2100 – but now several leading climate scientists believe that this could happen even before 2020.[4]

The acceleration of carbon dioxide (CO_2) emissions and its impact on sea-level can be seen by comparing data from the last 40 years or so with the past decade. According to measurements reviewed by the Intergovernmental Panel on Climate Change, the average rate of CO_2 emission into the atmosphere has increased from 1.4 parts per million (ppm) per year averaged over the period 1960–2005 to 1.9 ppm per year averaged over the period 1995–2005. Particularly worrying is the observation that the rate of sea-level rise has increased from 1.8 millimetres (mm) per year averaged over the period 1963–2003 to 3.1 mm per year over the 11-year period 1993–2003.[5]

Global warming is triggering several processes that are feeding back to amplify the original warming (see below). Unless the nations of this planet set and act on rigorous greenhouse targets, we are all facing potentially the worst environmental crisis of the 21st century[6] and an economic crisis on the scale of the Great Depression of the 1930s and the world wars of the 20th century.[7] Only a nuclear war could compete in terms of the scale of potential devastation.[8]

Solutions

Solutions to the threat of climate change do exist, but we have to separate the gold from the dross. My previous book *Greenhouse Solutions with Sustainable Energy*[9] assessed the various energy technologies that have been proposed for reducing greenhouse gas emissions. It found that coal power with *CO$_2$ capture and sequestration (CCS)*, sometimes misleadingly marketed as 'clean coal', is an unproven technological system that is unlikely to be commercially available before the 2020s. CCS merits development, but not at the expense of cleaner and safer technologies that are ready now.

Greenhouse Solutions also pointed out that nuclear power, based on existing 'burner' reactor technology, would become a significant greenhouse gas emitter within several decades, when *low-grade uranium ore* has to be mined and milled using *fossil fuels*. Although alternative reactor designs (such as the *fast breeder* and the *integral fast reactor*) could (in theory) overcome this problem, they too cannot be commercially available before the 2020s, possibly even 2030. Since nuclear power is a very slow technology to be deployed, it is not a short-term part of the solution and it may not even prove to be a long-term contributor.

The real solutions are staring us in the face. They are:

- efficient energy use;
- renewable sources of energy;
- natural gas, the least polluting fossil fuel, playing a transitional role while it lasts;
- improvements in urban public transport and better facilities for walking and cycling;
- an end to logging native forests;
- modifications to our lifestyles, including diet;
- modifications to agriculture to reduce methane emission from cattle and sheep and nitrous oxide emissions from soil, and to increase carbon sequestration in vegetation and soils.

These technologies and measures are ecologically sustainable and most of them are available now, although they may take some time to implement on a large scale.[10] Their roles are summarised in chapter 3. They are not the complete solution to global warming, but together they can achieve such dramatic reductions in greenhouse gas emissions before 2020 that they can buy us time to address the more difficult drivers of global warming: changing our economic system to a steady-state and stabilising our population. In the longer term, by 2050, Australia could achieve zero net greenhouse gas emissions.[11] If this could be achieved by Australia, the biggest per capita greenhouse gas emitter in the industrialised world, then it could be achieved by most other countries.[12]

However, new and improved technologies and measures are not implemented automatically, especially when they involve major changes to essential systems, such as energy demand and supply. To assist the transition, we need appropriate policies from government. Prior to 2009, the ecologically sustainable solutions were not favoured by the power-holders in the USA, the UK or Australia. Indeed, the governments of these countries have long-standing policies that favour fossil fuels and the nuclear industry.[13] To implement the solutions to the anthropogenic climate crisis, we need new government policies to build markets for *energy efficiency* and renewable energy, foster research and development, and encourage and assist households, workers, business and industry to make changes.

In the longer term, the barriers to ecologically sustainable solutions are not primarily technological or economic. Instead, they are political, institutional and cultural.

Barriers

Up to 2009, the world's biggest greenhouse gas polluters – the USA and China – and the biggest per capita greenhouse gas polluter of the industrialised world, Australia, have been doing little to cut their emissions. At best they are reducing slightly the respective rates of growth of their emissions. Only Europe is making a serious attempt to reach a significant short-term greenhouse target and, even there, we see indications of tokenism and backsliding.

Although public concern has been growing rapidly around the world, the majority of governments are dominated by vested interests, the industries that are the biggest greenhouse gas polluters: coal, oil, aluminium, steel, cement, motor vehicles, forestry and agriculture.

These vested interests have short-term time horizons and so do the politicians who support them. In their professional capacities, most of their CEOs do not appear to take seriously disasters that are likely to occur from 2020 and beyond. Neither do most of the politicians, whose concerns rarely extend beyond the next election.

The dominant economic system fosters societies based on endless growth in consumption; waste of energy, materials and land; and technologies that are dirty and appear to be cheap, but impose huge indirect costs on society and the planet. The economic system creates institutions and cultures that reinforce itself: employment that depends on unsustainable economic growth; and governments that subsidise the richest and most powerful industries and use taxpayers' money to bail them out when their speculations are unsuccessful.

Hope

Yet there is still hope. The economic power-holders and their representatives in government are just a tiny minority of the population. They maintain their dominance by misleading the public with simplistic notions, such as 'What's good for General Motors is good for the country',[14] and token gestures.

At this late stage in the process of global climate change, the principal hope of getting governments to facilitate deep cuts in emissions instead of token gestures is the rapidly growing social movement for climate action. That movement is making itself felt in Europe, North America, Australia and elsewhere. It needs resources, strategies and tactics to expand the knowledge of its members and to increase its influence on government and industry.

Before advancing further along the path of hope, some definitions are needed. My favourite definition of 'social movement' comes from the activist Bill Moyer,[15] who describes it as:

> collective actions in which the populace is alerted, educated and
> mobilized, sometimes over years and decades, to challenge the

power-holders and the whole society to redress social problems or grievances and restore critical social values

adding that

social movements are a powerful means for ordinary people to successfully create positive social change, particularly when the formal channels of democratic political participation are not working and obstinate powerful elites prevail.

In this book I call the social movement for action on global warming the climate action movement. This movement is composed of *climate movement organisations* (CMOs) and individuals. The different types of CMOs are identified in chapters 5 and 6. They include dedicated grass-roots climate action groups, large environmental and social justice NGOs for which climate action is one program, and some student, faith, trade union, business and professional groups.

Plenty of information is already available on the science and impacts of global climate change and the scientific and technological solutions. These resources come in the form of books, reports and websites (see 'Key readings and websites'). They cover the field at different levels of difficulty, from the popular to the scholarly. However, there is less information on policy options for governments and very little on strategies and tactics to guide members of the climate action movement. This book is intended for those of you who are sufficiently concerned about climate change to go beyond changing light bulbs, to join a CMO and to devote even an hour or three per week to this vital cause. The focus is on concerned citizens who live in a society with a system of government that purports to be democratic.

Who is a climate change activist?

Every reader of this book is potentially a climate change activist, if she or he is not already one. You are reading this book because you are concerned. You may already be campaigning as a member or supporter of an environmental, social justice, professional or faith group, working to reduce greenhouse gas emissions and stabilise our climate. You may have written a letter to a newspaper or participated in talkback radio on the issue. You may have arranged a meeting with your local member of parliament. Or you may simply have discussed the issue with family, friends, neighbours, work colleagues, or fellow members of your professional *organisation*, trade union, club or society.

If you are not already engaged in one or more of these campaign activities, then I encourage you to take some small steps while reading this book. If you have not already done so, you could join a CMO. A group can achieve more than the sum of actions of its individual members. Knowledge brings responsibility for action. With your knowledge and your actions, you can help to protect this planet for the present generation and all future generations. If you and others like you don't act to exert pressure within the democratic system, our governments will surely fail us.

Making democracy work

Most readers in the English-speaking world live in societies that are supposed to be democratic. These societies either believe or pay lip service to the idea of 'government of the people, by the people, for the people'.[16] In practice, some politicians take the view that all decisions should be left to the elected representatives and that the only role for the people in a democracy is to vote in an election once every three or four years. Large industries and businesses, professional organisations and other powerful groups encourage that limited view of democracy. Meanwhile, they lobby governments behind the scenes and disseminate material through the media to obtain policies to benefit themselves.[17] For example:

- the petroleum and motor vehicle industries lobby for tax deductions for petroleum exploration and for the use of company cars;
- the wealthy coal industry lobbies for approval of a new coal-mine or a new coal-fired power station, for huge government grants

and for taxpayer-funded infrastructure, such as ports, roads and railways for transporting coal;

- the aluminium industry lobbies for subsidised prices for its huge electricity purchases;
- the forestry industry lobbies for approval to log new areas of native forest;
- the livestock industry lobbies the government to market the country's beef and lamb overseas.

These lobbying activities – including direct meetings with ministers, other parliamentarians, political advisers and public officials – are backed up with political donations, media and advertising campaigns. These campaigns by vested interests bias the decisions of government and cause increases in greenhouse gas emissions.

Individual citizens who exercise their democratic right to redress the balance, carry little weight when approaching government or seeking media coverage for their concerns. However, citizens can greatly strengthen their influence by joining a CMO and becoming active within it. As awareness of the threat of global climate change grows, such groups are multiplying into a climate action movement that is local, national and international.

This book aims to assist you and other citizens to exercise your democratic rights, and educate yourselves about the issues and become climate action campaigners. First, we summarise the reasons why global warming is a serious and urgent issue.

Greenhouse gases cause global warming

If we are to be successful in our campaign to cut greenhouse gas emissions and stabilise our planet's climate, we must diagnose correctly the source of Earth's fever. The vast majority of climate scientists recognise that most of the global warming that has occurred since the Industrial Revolution is the result of human activities. These activities cause the emission of greenhouse gases.

Like the glass in a greenhouse, these invisible gases allow the life-giving rays of sunlight to pass unimpeded through the atmosphere. This warms the Earth and powers the process of *photosynthesis*, producing carbohydrates as stored solar energy. The warm Earth emits heat radiation that tries to escape the planetary bounds, but this infrared radiation is impeded on its way out by the greenhouse gases.

A balance exists between energy flowing into the Earth from the Sun, mostly as visible sunlight, and the energy flowing outwards from the Earth, mostly as invisible infrared radiation. As concentrations of greenhouse gases in the atmosphere increase, this balance is maintained, but something has to give: the Earth's temperature rises.

Before we humans interfered with the climate, natural concentrations of greenhouse gases in the atmosphere kept Earth's average temperature at about 14 degrees Celsius.[18] The principal natural greenhouse effect comes from water vapour, to which is added carbon dioxide (CO_2) and traces of methane and nitrous oxide. These greenhouse gases make possible the great abundance of life on Earth.

However, human activities, especially those carried out since the Industrial Revolution, have increased the concentrations of greenhouse gases in the atmosphere. The principal activities of concern are the combustion of fossil fuels – coal, oil and gas – to generate electricity and heat and to provide the motive power for transportation, while emitting CO_2 as a byproduct. Other important greenhouse gas-emitting activities are forest clearing, which results in the emission of CO_2 and methane, and agriculture, which emits lots of methane together with nitrous oxide and some CO_2. There are also industrial emissions, in addition to those involving energy, comprising CO_2 and some rare but potent greenhouse gases such as HCFCs and PFC. Fugitive emissions, which are leaks of methane and other greenhouse gases from fossil fuel production and transportation, must also be taken into account.[19]

In prehistoric times the Earth's climate varied enormously. Over billions of years, the Earth has experienced four major ice ages and long intervening periods when there were no polar ice-caps. The causes of these past variations were natural and are well understood by scientists. They involved regular oscillations in Earth's orbit around the Sun and in the Earth's tilt angle, all on different timescales. Changes in the amount of energy radiated by the Sun also played a role. Such natural effects can only explain a small fraction of the relatively large increase in Earth's average temperature of 0.75 degrees Celsius observed over the past century. They cannot explain any of the rapid increase that occurred towards the end of the 20th century.

The basic physics, that increasing greenhouse gas concentrations cause global warming, has been well established since the late 1950s and is not scientifically controversial.[20] Only the quantitative details, such as how much warming will result from a doubling of greenhouse gas concentrations, and the magnitude of the contribution of human activities, are subject to ranges of uncertainty and debate. The overwhelming evidence, that we humans are responsible for the major proportion of the industrial age global warming, is reviewed in the greenhouse science publications listed under 'Key readings and websites'.

The impacts of global climate change are already with us. As mentioned above, they include sea-level rise, droughts, heatwaves, bushfires, floods, loss of biodiversity and possibly an increase in the prevalence of intense cyclones/hurricanes.[21] These are neutral unemotional terms from the scientific literature that allow us to avoid recognising the pathway we have taken. 'Loss of biodiversity' really means 'the extermination of a large fraction of the species on the planet'. 'Sea-level rise' should be reframed too: it means 'the initiation of ice-sheet disintegration and sea-level rise, out of humanity's control, eventually eliminating coastal cities and historical sites, creating havoc, hundreds of millions of refugees, and impoverishing nations'.[22] Although it is scientifically difficult to attribute a single weather event to human-induced global warming, just as it is difficult to attribute a single case of cancer to smoking, the statistical evidence is very strong and growing. Even in a few individual events, such as the 35 000 premature deaths from the European heatwave of 2003, a scientific case has been made that global warming greatly increased the severity and impact of a 'normal' heatwave.[23] Similarly, global warming probably increased the severity of the bushfires that devastated Victoria (Australia) in February 2009,[24] and is increasing the likelihood of similar events around the world.

Why this is an urgent problem

Most scientists are cautious and try to avoid alarming people. The 2001 report of the Intergovernmental Panel on Climate Change (IPCC) showed that global warming is a very serious issue. However, it still conveyed a sense that climate change will occur gradually, with a very cautious treatment of the potential for sudden shocks or irreversible changes:

> Greenhouse gas forcing in the 21st century could set in motion large-scale, high-impact, non-linear, and potentially abrupt changes in physical and biological systems over the coming decades to millennia, with a wide range of associated likelihoods.[25]

To prepare the next (2007) IPCC report, scientific inputs had to be closed off in mid-2006. Even so, the 2007 report recognises that the previous report may have underestimated the magnitude and rate of some potential changes, for example:

> There is better understanding than in the TAR [Third Assessment Report of 2001] that the risk of additional contributions to sea-level rise from both the Greenland and possibly Antarctic ice sheets may be larger than projected by ice sheet models and could occur on century time scales.[26]

Since 2006, more and more scientific papers have been published providing evidence that several positive *feedback* or amplification processes for global warming are under way (see figure 1.1). These tell us that we must actually *reduce* the greenhouse gas concentrations in the atmosphere, below the present levels, and that time is of the essence for doing this. The IPCC (2007a) report only reviewed the effect of including *one* of these amplification processes in the modelling: increasing water vapour in the warming atmosphere. Now scientific concerns are mounting over at least *six* of these processes. These are discussed below.

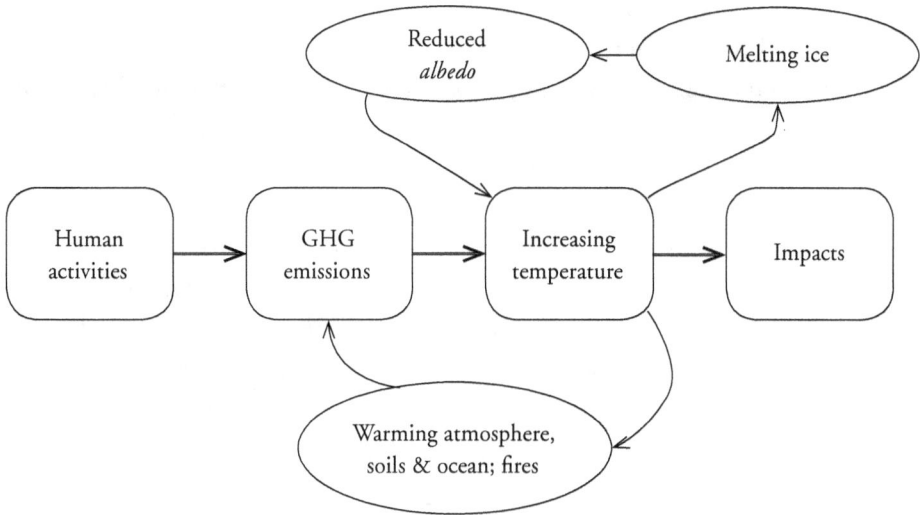

FIGURE 1.1 Positive feedback processes amplifying emissions

Arctic ice cap

From 1979 to 2008, the surface area of the Arctic ice cap has been declining at the average rate of 11.7 per cent per decade, as measured at its minimum in September.[27] The remaining ice, heated by warmer water underneath it, is thinning, increasing the probability that the ice cap will melt completely in Summer. A few climate scientists have calculated that this could occur as early as 2013, while others estimate 2030. As its ice cap shrinks, Earth reflects less sunlight and so it absorbs more sunlight, amplifying the existing warming.

Permafrost melting

As the Arctic Ocean warms, it in turn warms and melts the *permafrost*, which is frozen ground in Siberia, northern Canada, Alaska and in the Arctic sea-bed itself. As the permafrost melts, it releases trapped greenhouse gases, methane and CO_2, and so amplifies global warming.

Seawater warming

Warming of seawater over continental shelves and slopes may melt *hydrates* that currently hold large quantities of methane, thus amplifying warming.[28]

Wildfires

Global warming increases the prevalence and severity of wildfires, which release CO_2 from burning vegetation and peat accumulated over centuries, amplifying global warming. Climate scientist David Karoly argues that global warming contributed to the severity of the 2009 firestorms in south-east Australia.

Soil

As soil warms, it tends to release CO_2, amplifying global warming. On average, soil temperatures are warming as part of the general global warming process. At present soils are a major sink of CO_2, but if warming continues they could become a net source beyond 2050.[29]

Water vapour

As global warming proceeds, the atmosphere takes up more water vapour from oceans, lakes and rivers. Since water vapour is a greenhouse gas, this amplifies global warming. Some of the additional water vapour will form clouds. High clouds reflect sunlight, thus introducing a cooling effect, while low clouds absorb more infrared radiation from the Earth, thus producing a warming effect. Although the net effect of the additional clouds is uncertain, the effect of the water vapour that is not in the form of clouds is definitely one of warming and is included in some recent climate models.[30]

Oceans

At least two amplification processes for global warming may occur in the oceans. Firstly, as the oceans warm, they absorb less CO_2 from the atmosphere, thus leaving more in the atmosphere where it amplifies warming.[31] Secondly, some of the CO_2 that is absorbed by the oceans forms carbonic acid, thus making the water more acidic. The acid attacks and dissolves the armour of sea creatures, which is made of calcium carbonate, destroying biodiversity and emitting CO_2, some of which may escape into the atmosphere, amplifying warming.[32]

CO_2 fertilisation

There is also a negative feedback effect, in addition to the reflection of sunlight from high clouds. CO_2 increases the rate of growth of plants. However, this is constrained by the limited availability of additional water and nutrients in many regions and so this effect is nowadays regarded as less significant than previously.

Summing up the amplification effects

It may already too late to save the Arctic ice cap. An indirect effect of its melting, warming of the permafrost and the Arctic Ocean, is of particular concern. The quantities of methane and CO_2 stored there are very large. If all these greenhouse gases were eventually released, they could greatly increase the existing *CO_2-equivalent* concentration in the atmosphere.

Fortunately, the other amplification processes still appear to be at an early stage of development. The only way to stop such positive feedbacks is to cut them off while they are still tiny perturbations on the existing trend. Once they become significant in size, it will be too late. If all the above positive feedbacks take off, the Earth's climate will change dramatically and irreversibly. It could warm by 6–10 degrees Celsius on average, with even greater increases at the poles. This in turn could melt much of the ice on Greenland and West Antarctica, lifting sea-levels by tens of metres within the 21st century.[33] Almost all coastal cities would be inundated.

Why climate action is an ethical imperative

The climate crisis and its response strategies are a universal problem, involving environmental protection, social justice, economic impacts, governance and ethics. The latter aspect has been downplayed by the big greenhouse gas polluters and their representatives in government, because (I suspect) they understand that their ethical position is very weak. Unfortunately the media and a large proportion of the population have taken their cue from the vested interests and largely ignored ethics. Instead they have focused on the potential economic impacts of greenhouse *mitigation* on these vested interests. These ramifications are dressed up as adverse impacts on the whole economy.[34] People who find the scientific and economic issues of climate change too complicated to follow can readily understand the ethical issues. We must continually bring before the public the following key ethical issues:

- In the words of Australian Greens Senator Christine Milne, 'do we value our children as much as ourselves?'[35]

- The most damaging impacts of climate change are likely to be on less developed countries, which have the least responsibility for the problem and the least capacity to alleviate impacts. Therefore, the developed countries have a moral responsibility to take the lead in reducing greenhouse gas emissions and to compensate the less developed countries.

- Mitigation policies involving pricing will have the greatest economic impacts on low-income earners, whether they live in less developed or highly developed countries. Hence strategies for assisting low-income earners to reduce their emissions and to compensate them for higher energy prices must be implemented at the outset.

- The quantity of greenhouse gases that can still be emitted into the atmosphere, with a high probability of avoiding serious damage to the planet and its people, is now very limited. The rich countries have built their wealth on industries and processes high in greenhouse gas emissions. Yet a similar development pathway cannot be followed by less developed countries without impacts that could destroy human civilisation and much biodiversity. This implies that any global agreement must require both rich and poor countries to change their economic/industrial structures, with the rich assisting the poor. This must involve a large transfer of wealth from rich to poor countries. Thus the moral requirement of the first dot point is reinforced by the practical requirement of the third dot point.

The process of *Contraction and Convergence,* proposed by the Global Commons Institute,[36] is consistent with the resolution of the above international ethical issues. The process is that the developed countries decrease their per capita emissions, while the developing countries increase theirs, until all countries have the same average per capita emissions at a level that has low climate risk. In practice, the greed of the rich is likely to be a huge barrier to this solution.

Plan of the book

Climate change campaigners need to know more than strategies and tactics. To be effective in convincing people in all walks of life, you must also be able to demystify the fallacies disseminated by vested interests, to explain the potentials of various clean energy technologies, and argue the case for the government policies needed to achieve big reductions in emissions. Each of the following chapters is designed to inform and assist you in one of these essential skills.

Chapter 2 is designed to give you a basic understanding of the politics (in the broadest sense) of climate action. It first identifies the driving forces of greenhouse gas emissions: population, per capita consumption and technology choice. Underlying these forces are powerful vested interests, which exert strong influences upon government policies and community attitudes. They promote industries, businesses, technologies, policies and individual lifestyles that foster ever-increasing greenhouse gas emissions. An important aspect of their influence is their dissemination of myths and fallacies about *sustainable development* pathways in general and renewable energy in particular. Under considerable pressure from these vested interests, governments tend to implement token and trivial responses to the greenhouse challenge, surrounding these actions with 'spin' that attempts to cloak them in substance. This chapter is designed to assist concerned citizens and climate action campaigners to demystify the myths, fallacies and spin and to identify the real issues that need urgent attention.

Chapter 3 reviews the technologies that can achieve deep cuts in greenhouse gas emissions, especially from energy and transport. It also points out that technological improvements, while necessary and valuable, are not sufficient to solve the greenhouse problem completely. We must also address the other driving forces: population growth and per capita consumption. Underlying the three driving forces are greed, unjust and unsustainable economic systems, inappropriate systems of governance, and cultural factors.

Chapter 4 sets out the key policies that concerned citizens and campaigners must extract from all levels of government to achieve large, lasting and ecologically sustainable reductions in greenhouse gas emissions. Very few of these policies are being implemented to a significant degree anywhere in the world. The gap between needed and existing policies cannot be closed simply by communicating the facts to the power-holders and appealing to their integrity, altruism and public spirit. Unfortunately many of them have been captured by the vested interests.

The principal way of countering the influence of vested interests is to build a strong and committed nonviolent social movement. Chapter 5 develops broad strategies for the climate action movement. The chapter presents the case that the concentrated political power of the big greenhouse gas polluters can and must be matched by motivating and organising concerned citizens into a climate action movement that exercises countervailing power. To be effective, the climate action movement must become well organised, so that it can campaign at local, state, national and international levels. The *strategy* of a nonviolent campaign needs careful development and planning, using tools such as *SWOT analysis* for a minor issue and the Midwest Academy Strategy Chart for a major one. This chapter also introduces a strategic framework for understanding the dynamics of major campaigns by social movements, known as the Movement Action Plan. This presents a citizens' campaign as renewal of participatory democracy. By drawing upon the successes of the civil rights and the anti-nuclear power movements in the USA and several other social movements, it offers campaigners hope and practical guidance on strategies.

The next step is to understand and evaluate CMOs and their tactics, actual and potential. Chapter 6 identifies the different types of CMO and includes case studies of such groups in three countries: the USA, the UK and Australia. These studies identify the kinds of action being carried out, and the gaps that must still be filled. Drawing upon both the scholarly and practical literature on social movements, the chapter sets out some of the specific tactics that can be adopted or further developed by CMOs. A wide range of tactics and tools is needed: lobbying decision-makers; negotiating with and building alliances with influential organisations; networking with other climate movement organisations; educational activities; media and other communications; nonviolent direct action; legal action; and setting up alternatives.

The ultimate goal of such actions is twofold. Firstly we have to convince the majority of the population that strong actions are needed urgently to gain deep cuts in greenhouse gas emissions. Secondly we need to exert overwhelming pressure upon governments and businesses to implement effective policies without further delays.

GREENHOUSE MAFIA AND THEIR FALLACIES

Mining and burning dirty coal; exploring and drilling for oil and gas; processing minerals; smelting aluminium and other metals; logging native forests; farming cattle and sheep; and manufacturing, selling, fuelling and maintaining motor vehicles. There's a great deal of money being made by producing greenhouse gas emissions. So naturally there will be fierce resistance when scientists and the community say that emissions must be reduced. Vested interests (called the *Greenhouse Mafia* in Australia) and others who oppose change towards a genuinely sustainable society spread fallacies about greenhouse solutions (some might even go so far as to call them lies or outright deceptions). So the first step for you as a concerned citizen and climate action campaigner is to be able to understand and refute these fallacies. To assist this process, I first elucidate the driving forces of human-induced greenhouse gas emissions and then identify the vested interests feeding these forces. With that groundwork performed, I name and refute the principal fallacies about greenhouse solutions.

Driving forces of emissions

Emissions are determined by a combination of three factors: population, consumption per person (sometimes called '*affluence*') and technology choice. For simplicity, let's consider *carbon emissions C* resulting from energy generation E and population P. In this case consumption per person is simply energy use per person (E/P) and technology choice can be measured by carbon emissions per unit of energy use (C/E). Then we can disaggregate carbon emissions into three factors:

$$C = Population \times Consumption\ per\ person \times Technology\ impact$$

$$= P \times (E/P) \times (C/E)$$

Clearly this relationship is identically true, because we can cancel the *P*s and *E*s to obtain the identity $C = C$. If we double any one of the factors on the right-hand-side – population, consumption per person or technology impact – then carbon emissions are doubled. If we double all three factors, then carbon emissions are multiplied by 2 x 2 x 2 = 8.

Guided by this relationship, we can see that total emissions from the USA are very high because of very high consumption per person, high population (306 million in 2009) and a quite a high proportion of fossil fuel in the energy mix. Total emissions from China are very high, despite low average consumption per person, because of a very high population (about 1300 million) and a very high proportion of fossil fuel, especially coal, in the energy mix. Australia's total emissions are much lower than those of the USA and China because of its relatively low population, but Australia's unenviable record for record-breaking per capita emissions results primarily from its very high use of coal for electricity generation.[1]

Out of the collective 'we' who are responsible for greenhouse gas emissions, the principal culprits are large corporations (discussed below) and individuals who are affluent, in the sense of having high consumption per person, and are living in a society with *greenhouse-intensive* 'technologies', such as burning fossil fuels, energy inefficient buildings, logging native forests, and eating cattle and sheep. Nevertheless, we must be careful about blaming individuals, because most have limited control over their emissions, as discussed in the refutation of fallacy 1 below.

Large corporations are powerful forces in shaping the economy by lobbying governments and fostering consumer demand through advertising. They encourage growth in consumption per person by pandering to human greed and other weaknesses. They lobby for government policies to increase population in order to provide a cheap pool of labour and to boost specific industries such as housing/property. They also support the dirtiest technologies in terms of greenhouse gas emissions, because these are generally the cheapest. Indeed, they generally oppose measures to include the environmental and health costs of products in their prices.

Underlying the three drivers of emissions – population growth,

growth in per capita consumption ('affluence') and dirty technology – are of course greed, culture (including consumerism), an economic system that encourages growth and speculation, and insecurity resulting from poor governance and the economic system. Although we cannot eliminate greed as a basic human property, with sensible policies we can change the culture and economic system that give it encouragement, support and expression on national and international scales. With the right policies, we can make it easy for corporations and individuals to do the right thing.

Vested interests – Greenhouse Mafia

The situations in three countries – the USA, the UK and Australia – illustrate the influence of vested interests in greenhouse gas pollution on government decisions.

Vested interests in the USA

A major business/industry lobby group that influenced the US government against climate action was the Global Climate Coalition, formed in 1989. Prior to 1997, its members included the Aluminum Association, British Petroleum, DaimlerChrysler, DuPont, ExxonMobil, Ford, General Motors, National Mining Association, Shell, Texaco, and the US Chamber of Commerce. Its techniques included lobbying US politicians and delegates at international fora, public attacks on the credibility of the Intergovernmental Panel on Climate Change, advertising campaigns and the preparation of glossy reports and videos. As a result of the activities of the Global Climate Coalition and other US groups representing vested interests, the Senate unanimously passed a resolution that the United States should not be a signatory to any protocol that did not include binding targets and timetables for developing as well as industrialised nations or 'would result in serious harm to the economy of the United States'. This resolution was passed in July 1997, after the Kyoto Protocol had been negotiated and a penultimate draft produced, but before it was finalised.[2]

By 1997, due to growing scientific and public consensus certain members of this coalition realised that they had placed themselves in an untenable situation. Increasingly they were being recognised as self-serving anti-environment groups, and this could cause great damage to their public images. One by one, the coalition members followed the lead of BP and publicly abandoned it. In early 2002, when President George W Bush had indicated that he would reject the Kyoto Protocol, the Global Climate Coalition declared itself 'deactivated', explaining that it 'has served its purpose by contributing to a new national approach to global warming'.[3]

Under President George W Bush, the White House was criticised for censoring reports from government agencies on climate change research. In particular, the *New York Times* reported that a White House official, Mr Philip Cooney, repeatedly removed or adjusted descriptions of climate research that had already been approved by government scientists. Before working at the White House, Mr Cooney was a 'climate team leader' and lobbyist at the American Petroleum Institute, the largest oil industry trade group.[4]

Vested interests in Australia

In Australia, both the previous Howard Liberal-National Coalition Government (1996–2007) and the present Rudd Labor Government (2007–) have been strongly influenced by the big greenhouse gas-emitting industries to go slow on climate action.

One source of evidence supporting this statement is a set of leaked notes from a meeting on 6 May 2004 of the Lower Emissions Technology Advisory Group, which comprises CEOs of the major fossil fuel producing and consuming corporations. The meeting, which was addressed by the then prime minister John Howard and the then industry minister Ian Macfarlane, discussed ways and means of limiting the growth of the renewable energy industry and thus protecting the fossil-fuel based industries.[5]

This apparent collusion between the Howard Government and the big greenhouse gas polluters was subsequently confirmed by Guy

Pearse.[6] As a member of the Liberal Party and a former ministerial adviser in the Howard Coalition Government, Pearse was able to obtain frank interviews with the captains of these polluting industries for his PhD thesis. He reports that they boasted to him that they, the self-styled 'Greenhouse Mafia', were responsible for writing government policy on greenhouse response.

Additional grounds for concern that the Howard Government was attempting to suppress renewable energy comes from an apparently concerted government campaign against wind power, the least-costly of the new renewable sources of electricity. In this campaign, verbal attacks on wind power were uttered by at least five ministers, including the prime minister, within a period of about one year.[7]

The Rudd Labor Government won office in November 2007, partly because of its promises to expand renewable energy in Australia.[8] As of May 2009, no leaks or whistleblowers have revealed collusion with the big greenhouse gas emitters *per se*, though clearly emitters are having a strong say in the development of government legislation.[9] The new government's failure to implement its main election promises – by not making the necessary financial allocations to renewable energy in the May 2008 budget and its failure to set up the appropriate institutions in 2008 – are indications that the change of government has not necessarily brought a significant change of policy. Specifically, apart from the symbolic gesture of ratifying the Kyoto Protocol, the new government has delayed implementation of all of its principal election promises to support renewable energy. From a comparison of the November 2007 election promises[10] with the 2008 budget papers,[11] the following concerns arise.

1 Renewable Energy Fund

Before the election, Labor promised to create a $500 million Renewable Energy Fund[12] to be allocated over seven years for the development and deployment of renewable energy. But in its first budget in May 2008, Labor allocated nothing to renewable energy (apart from geothermal drilling) from this fund. However, it did allocate funding to the coal industry from its so-called Clean Coal Fund, also for $500 million.

2 Energy Innovation Fund

Before the election, Labor promised a $150 million Energy Innovation Fund[13] for renewable energy research. From this fund $50 million would be allocated to solar thermal energy research and $50 million to solar photovoltaic electricity research.[14] But in its first budget, the new Labor Government allocated zero research funding in 2008–09 to solar, wind and *bioenergy*. As of March 2009, it is still in the process of setting up the fund and the associated Solar Institute.

3 Renewable Energy Target

Before the November 2007 federal election, Labor promised to expand the *Mandatory Renewable Energy Target* (MRET) to 20 per cent of electricity by 2020. This means increasing the existing contribution from renewable energy – about 8.5 per cent of electricity or 15 TWh per year, mostly hydro – to about 60 TWh per year. However, after attaining government, Labor delayed implementation unnecessarily[15] by handing over the process to the Council of Australian Governments (COAG).[16] According to the government's White Paper, implementation will only take place on 1 January 2010, over two years since the government was elected.[17] Meanwhile the new government set up the Wilkins Inquiry to pronounce on whether such '*complementary measures*' as MRET were necessary once an *emissions trading scheme* is operational. Wilkins reported around July 2008, but the government has stated that it would never make this report public.

4 Subsidies for residential solar

In the May 2008 budget, the Labor Government introduced a means test of $100 000 gross household income to limit eligibility to the $8000 rebate for residential solar electricity. This was a weakening of an election promise. At the same time the government proposed to replace the rebate with a *feed-in tariff*, handing the proposal over to COAG, a sure recipe for delay. After considerable public debate throughout 2008, the government announced in early 2009 that on 1 July 2009 it would discontinue the rebate and not implement a feed-in tariff to replace it. Instead it would increase the number of 'solar credits' (renewable energy certificates) allocated to residential solar PV electricity under the new *Renewable Energy Target* by a factor of five. In this way, the government diminished support for residential renewable electricity, transferred the costs from itself to electricity consumers and in effect reduced the size of the Renewable Energy Target.

5 Role of Garnaut Review

Before the 2007 election, Labor announced that it would base its emissions trading scheme on the recommendations of the Garnaut Climate Change Review. However, in February 2008, when Garnaut published an interim report expressing the need for very strong action, the Minister for Climate Change backtracked, announcing that the Garnaut Review would be just one of several inputs to its policy.[18] The government's White Paper on climate change[19] ignores or waters down several of the key recommendations of the final report of the Garnaut Review.[20] As Garnaut points out, for an emissions trading scheme to be effective, it is essential to auction all permits and to use the revenue raised to assist low-income and other households to reduce their emissions and to fund new infrastructure. Garnaut also pointed out that paying 'compensation' to the largest greenhouse gas polluters would undermine the scheme. In its White Paper, the government attempts to placate the vested interests in greenhouse pollution by offering the coal-fired electricity generators free emission permits valued at $3.9 billion in the first five years of operation of the scheme and the emissions-intensive trade-exposed industries with initially 25 per cent of all permits free of charge, increasing to a maximum of 45 per cent by 2020. The so-called Australian Carbon Pollution Reduction Scheme is discussed further in chapter 4.

Delayed election promises

The first four of these were unconditional election promises that should not have been delayed until the Garnaut Climate Change Review and the government's White Paper on climate change had been published. Given the political will, they could all have been implemented in the first six months after the government gained office. The government's explanation for the delays is that they were needed for setting up administration. This is hard to believe. MRET already existed under the Howard Government, so the scheme could have been expanded immediately with the constraint that projects could not benefit from both the federal and state schemes. Furthermore, administration has not delayed funding for so-called 'clean coal'. The most likely explanation for the delays is pressure from the Greenhouse Mafia. While it is possible that all the 2007 promises will be implemented in some form in the near future, the long delay indicates that the government is unwilling to treat climate change as an urgent issue and give strong support for renewable energy, despite its rhetoric. The need for a growing climate action movement is clear.

If you have any lingering doubt that the Rudd Government has been captured by the Greenhouse Mafia like the previous Howard Government, inspect the membership of the High Level Consultative Committee for Australia's 2009 Energy White Paper at <http://www.ret.gov.au/energy/facts/white_paper/Pages/default.aspx>. The business representatives come from coal, oil, gas and uranium. There are no specific representatives for renewable energy or energy efficiency. Clearly, the result of the Energy White Paper process is predetermined.

Not surprisingly, a public opinion poll conducted by Auspoll for the Climate Institute in October 2008 found that, of the 1000 people polled, 58 per cent could not distinguish between the major political parties' climate policies. Only 28 per cent of Australians think Labor is the 'party better able to handle climate change'. In October 2008, Labor was at its lowest polling on this question since February 2007, down from 34 per cent in July 2008. The Coalition had risen from 9 per cent in July to 14 per cent in October. Public concern about climate change fell marginally in the six-month period since March 2008, but remained high, down from 89 per cent to 82 per cent, despite the financial crisis.[21]

Vested interests in the UK

In the UK, the coal industry and coal trade unions are much smaller and weaker than in Australia or the USA, and British Petroleum is now calling itself Beyond Petroleum. So the vested interests that may be influencing the British Government are less obvious. Overt anti-climate action groups provide superficial arguments and unconvincing positions, even by the low standards of the anti-climate action movement. The presence of stronger vested interests behind the scenes can be inferred by two of the UK Government's recent decisions on energy supply, both announced in January 2008:

- The decision to build a new conventional coal-fired power station at Kingsnorth, despite the fact that the original proposal required it to have carbon capture and storage (CCS) before it would be allowed to operate.

- The sudden reversal of the government's 2003 policy not to build new nuclear power stations.[22]

In both cases, it may be relevant that UK's electricity generation and retail system is now mostly owned and controlled by French and Germany utilities – EDF, E.ON and RWE – so it is possible that they are shaping the policies the UK government adopts. Indeed Greenpeace UK believes that the driving force behind the decision to allow Kingsnorth to go ahead without CCS was explicitly the giant European utility E.ON. Under Freedom of Information, Greenpeace UK obtained copies of emails between the Department of Business and E.ON. On 16 January 2008, an email from E.ON demanded that the department radically alter the conditions that it had placed on building Britain's first new coal-fired power station in 30 years. Originally the government had required that the proposed new Kingsnorth power station be fitted with CCS before commencing operation.

E.ON's email admitted that CCS technology is not currently viable at any scale and demanded that E.ON be allowed to build a conventional dirty power station. Just six minutes after E.ON's email was sent, a senior member of the department replied: 'Thanks. I won't include [the previous conditions].'[23]

In 2003 the British White Paper on Energy stated that 'the current economics of nuclear power make it an unattractive option for new generating capacity'.[24] Yet in January 2008, the government announced that it would facilitate the construction of new nuclear power stations. In a sentence buried in the main body of the report (Paragraph 3.52), but not in the executive summary, it even offered an economic bailout:

> Operators are responsible for decommissioning and waste management costs. If the protections we are putting in place through the Energy Bill prove insufficient, in extreme circumstances the Government may be called upon to meet the costs of ensuring the protection of the public and the environment.[25]

EDF (Électricité de France), the world's biggest operator of nuclear power stations, is in the process of purchasing British Energy, the owner of most of the UK's nuclear power stations. This purchase would make EDF the largest electricity generator in the UK, giving it enormous political power over energy policy decisions by the UK government. This purchase and the recent sale of the Sellafield nuclear *reprocessing* facility to a US-based private consortium would make it more difficult to obtain information under Freedom of Information.

Campaigns by vested interests

The vested interests in fossil fuels and nuclear energy are largely responsible for two major campaigns against greenhouse gas reductions:

- the campaign to sow doubt about greenhouse science;
- the campaign to delay, divert and undermine greenhouse mitigation.

Both these campaigns have been conducted by disseminating fallacies. We'll discuss the main fallacies below.

Fallacies about greenhouse science

For decades the big greenhouse gas emitting industries have disseminated myths, fallacies, 'spin' and outright lies about the science of global warming from the human-induced greenhouse effect. They have deliberately sown doubt and confusion in the minds of politicians, journalists and members of the public at large. According to the Royal Society, the organisation of eminent British scientists, the oil company ExxonMobil has paid millions of dollars per year to groups that misrepresent greenhouse science.[26] However, by 2006, greenhouse science, supported by observations of widespread and growing climatic impacts and popularised by Al Gore's film and book *An Inconvenient Truth*,[27] triumphed at last in the minds of the vast majority of Europeans and Australians, who now accept that global warming is a real, major and urgent issue.

For greenhouse campaigners in these countries, refuting again and again the fallacies that vested interests have disseminated is a waste of time. That task can be safely left to climate scientists, some of whom are becoming more outspoken in the face of misrepresentations of the anti-climate action organisations and individuals. However, the deniers of climate science are stronger in the USA. The key fallacies about greenhouse science are addressed in chapter 2 of *Greenhouse Solutions with Sustainable Energy* and in more scientific detail on the websites of Real Climate and Stephen Schneider.[28] This section addresses one recent false claim that global warming ended in 1998 and that Earth has now entered a period of cooling.

Fallacy: Earth has been cooling since 1998

The global average temperature fluctuates from year to year, because it is subject to several natural short-term influences superposed on the long-term human-induced global warming trend. For instance, in the decade 1989–98 inclusive, there was a strong increase in temperature, but not every year was warmer than its preceding year. Natural fluctuations were responsible for the year-to-year variations. In the decade 1998–2007 inclusive, the two warmest years were 1998 and 2005, in that order. However, it is misleading to argue from this that Earth has entered a period of cooling. Comparing 1998 and 2005 is like comparing apples and oranges: 1998 was the warmest *El Niño* year on record, while 2005 was a *La Niña* year. El Niño is a natural climate phenomenon in which the eastern tropical Pacific Ocean and hence the rest of the world is warmer than average, while La Niña makes the eastern Pacific and the rest of the world cooler than average. If 2005 had been a strong El Niño year, then it would have been the warmest year ever experienced by humans. But picking out a single year, whether it be 1998 or 2005, is bad science, because it does not give enough data to establish a trend in climate. For this reason, climate scientists average temperatures over minimum periods of a decade to identify climate trends. When they do this, they find that the decade 1998–2007 was the warmest in recorded history measured by thermometers.

So, in short, the answer is no, the Earth is not cooling.

Fallacies about greenhouse solutions

Having lost the debate about greenhouse science, the vested interests, with the assistance of politicians and uncritical journalists, are disseminating misinformation and confusion about potential solutions to global warming. In order to assist you to counter such misinformation, this section is devoted to mythbusting. It provides brief refutations to 17 fallacies about greenhouse solutions. These fallacies take different approaches:

- some claim that action would be too expensive for the economy, in terms of impact on gross domestic product (GDP) and jobs;
- others seek delays in action;
- others attempt to divert action towards technologies (such as

nuclear power) and policies that would be ineffective in reducing emissions;

- while others undermine the technologies and policies that offer the principal solutions, energy efficiency and renewable energy.

As in the case of climate science, most of the fallacies about climate mitigation are based on a few drops of truth in a cup filled with lies, exaggerations and misleading frameworks.

FALLACY 1

All we need to do is change individual's behaviour

The fallacy runs along the lines that governments and corporations are composed of individuals and therefore all the climate action movement has to do is to convince the individual members of these organisations and the problem is solved.

Response: The fallacy in this argument is its failure to recognise that organisations, whether they be government departments or business corporations, have structures and goals that generally make them behave in quite different ways from their individual members. For example, public companies are legally responsible to their shareholders, while private companies are legally responsible to their owners. In the Westminster system, government departments are responsible to their respective ministers, while in the US system secretaries and other senior managers of government departments are appointed by and are responsible to the president.

Even when the majority of members of a public or private company or a government department support strong greenhouse action, the structures of these organisations and their goals generally preclude any change. The most prevalent organisational structure is that of the bureaucracy, a hierarchical system in which individual workers are interchangeable. This is an excellent system for carrying out routine operations efficiently under central direction, but it is very resistant to change.

In a few rare cases, the most powerful person in a bureaucracy has initiated change in the public interest, either as the result of external pressure such as a consumer boycott, or rational argument, or revelation. A famous example is Ray Anderson, founder and chairman of Interface, one of the world's largest floor covering companies, who experienced a revelation, a 'spear in the chest' as he describes it, and made a 'mid-course correction' to take his company onto a sustainable development pathway.[29] This inspiring story is the exception that proves the rule. Ray Anderson was by far the most powerful member of his organisation.

Another statement of the fallacy that individual action is sufficient is based on a simplistic story called *The Hundredth Monkey*,[30] which is critiqued in *Greenhouse Solutions with Sustainable Energy*, pp 330–31. Human society (and probably monkey society) is more complicated than assumed in this story.

Some organisational structures, notably cooperatives, are more responsive to the views of their members than companies. A century ago, cooperatives were much more prevalent in Western countries than they are today. One inspiring example of the success of worker cooperatives is the transformation of the formerly impoverished region around Mondragon, in the Basque region of Spain, by the formation of a network of producer and consumer cooperatives.[31] Although nowadays corporations rule most of the economies of the world,[32] cooperatives are undergoing a renaissance and are proving more resilient than some corporations in the face of the global economic crisis.

Not only does the fallacy fail to recognise the inability of individual action to change corporations, but it also ignores the lack of power of individuals to change laws, infrastructure and the economic system. Governments collect a large fraction of national wealth in the form of taxation and governments decide how to spend that wealth. They make the laws, which assist some actions and impede others. Will most transport funding go to roads or to public transport, cycleways and footpaths? Is a new power station really necessary and, if so, will it be fuelled by coal, natural gas or renewable energy? Will all houses and commercial buildings meet energy efficiency standards? Will our nation have a population policy? Will native forests continue to be logged? Will farmers be allowed to clear their land and, if not, will they be compensated? Where will a new business district or shopping mall of a city be located and will it be accessible by public transport? Will a proposal for a new aluminium smelter be approved and will its electricity price be subsidised, as is the usual practice?

In terms of greenhouse gas emissions, government decisions swamp those of individual people and individual households. Furthermore, large corporations are the ones that influence all of these decisions by government. Thus individual actions are constrained by much larger forces. This is even true on the domestic scale. If we live in rented housing, we cannot install insulation or a solar hot water system; that is the prerogative of the landlord. In the face of these formidable barriers, we have three principal responsibilities for reducing greenhouse gas emissions:

- The first and key responsibility is, as citizens, to exercise our democratic rights *collectively* to demand that our governments take action by implementing appropriate laws, regulations and standards, pricing and funding policies, education and information, industry policies, institutional change and population policies. As we shall see in subsequent chapters, our democratic responsibilities go far beyond voting as individuals once every three or four years. The collective actions of citizens cooperating are needed to change government policies. In the words of Helen Keller, 'Alone we can do so little. Together we can do so much.'

- The second is, as consumers, to apply collective pressures to industry and business to supply ecologically sustainable and socially just products and services.

- The third, and only the third, is to do what we reasonably can as individuals and households to reduce our own direct emissions from electricity, heating and cooling, transportation, food, etc.

With strong and appropriate government action to shape markets, all three types of action would become easy.

None of the foregoing arguments should be used to discourage individual actions. Making our own homes more energy efficient, buying Green Power and driving less are empowering and healthy. These actions set an example to friends and neighbours. To have credibility in asking others to act, we must demonstrate commitment to reducing greenhouse gas emissions in our own lives. Energetic and highly committed individuals can also play a vital role as *champions* of particular climate action campaigns and leaders of climate action groups. Individual actions are necessary and valuable, but not sufficient to meet the challenge of climate change. To gain effective climate change policies from government and business, pressure from a climate action movement, composed of a multitude of climate movement organisations (CMOs), is needed.

FALLACY 2

Unless big emitters take action, we should do nothing

Since most countries each emit less than two per cent of global greenhouse gas emissions, the fallacy goes that reducing their emissions would have negligible international impact. Therefore, they should wait for the big emitters, USA and China, to take action.

Response: Fortunately for our planet, this notion has been rejected by the European Union (EU), which is building for its members a coordinated program with EU targets and timetable. However, this fallacious argument is prevalent among groups resisting action in the UK, Canada, Japan and Australia. In particular, the international *political* impact of Australia changing its stance could be large, since Australia is the industrialised world's biggest per capita greenhouse gas emitter. If Australia supported strong short-term targets for the next international agreement and took substantial action to reduce its own emissions by 2020, it would be clear that most other countries could achieve the same or better targets. It would add further pressure on the USA to take stronger action. The benefits to the world of leadership from countries with low greenhouse gas emissions should not be underestimated.

Furthermore, substantial action by the developed countries, especially the USA and Australia, is a necessary pre-condition for bringing developing countries such as China and India into an international agreement with targets. This was underlined by a senior Chinese climate adviser when he stated that Australia would derail global climate talks if its 2020 greenhouse target was less than a 25 per cent cut in emissions.[33]

FALLACY 3

Climate change is unstoppable

The naysayers tell us that climate change is unstoppable. Therefore it's better to avoid the costs of mitigation and spend the money saved on *adaptation*.

Response: In the specialised language of climate science, 'mitigation' means reducing greenhouse gas emissions, while 'adaptation' means reducing the impacts of climate change without addressing emissions.

One of the few people to articulate fallacy 3 publicly is Gary Johns, a former Labor Government minister in Australia who became a senior fellow with the Institute of Public Affairs, a right-wing 'think-tank' that is funded by the resource industries among others.[34] Several questionable assumptions underlie this position.

The first is that action by Australia to reduce its emissions will not influence other countries to reduce their emissions. This assumption is refuted in fallacy 2.

The second is that expenditure on mitigation will be ineffective and so will not reduce the cost of adaptation. The refutation runs as follows: provided the global mitigation effort is sufficient to avoid a catastrophic tipping point, such as one leading to the melting of all ice on Earth, some mitigation is better than no mitigation. Climate modeling reviewed by the IPCC shows that the adverse impacts of a

2 degree Celsius average global warming are likely to be less than those of a 3 degree Celsius warming, etc. If we assume the worst, then it will become a self-fulfilling prophecy. Respect for our children and grandchildren demands that we reject the selfish position of 'après moi le deluge' (literally, 'after me, the flood', or more loosely 'after I'm gone, the future doesn't matter').

The third questionable assumption is that the cost of adaptation will be much less than the cost of mitigation. Mr Johns' article suggests naïvely that adaptation is simply a matter of building a few more dams to store water. But, if climate change brings permanent drought, as has already happened in south-west Australia and appears to be happening in the Murray-Darling river system and south-east Australia, then no amount of dams will solve the problem. In reality, drought and rising temperatures are threatening Australia with the loss of a large part of its agriculture and tourism industries before 2020 and are feeding terrible bushfires. Rising sea-levels could severely damage almost all its cities before 2100. In Asia, almost all the major rivers are likely to dry up as the glaciers that feed them on the Tibet-Qinghai plateau disappear, leaving a billion people without sufficient water. Two metres of sea-level rise could be devastating to the Netherlands, the UK, China, Bangladesh and many other countries. On a global scale, the fallacy is refuted by the Stern Review (see fallacy 11). If we don't mitigate, the impacts of climate change will be more severe and adaptation much more expensive.

The main problem in slowing human-induced climate change is that very few governments are willing to take the lead. Yet the situation, which is known as the Environmental Prisoners' Dilemma, does have a solution, albeit a challenging one.[35] If every nation waits for the rest of the world to take action, we will all lose. The best course of action for escaping the dilemma is for each nation to take strong action as soon as possible, setting an example that encourages other nations to join in. Then we can all win or at least diminish the damage.

CCS is the principal greenhouse solution

This fallacy states that coal power with CO_2 capture and sequestration (CCS) is the principal greenhouse solution.

Response: Although a few components of this proposed solution exist, coal power with CCS is an unproven technological system. Although pilot plants will be built before 2020 as governments pour in vast amounts of money, this will still be a long way from full-scale commercial production with a high confidence in safety. In 2008, the US Department of Energy terminated its funding for FutureGen, its principal project to build a coal-fired power station with CCS, because of rapidly escalating costs. The risks of CO_2 escapes from underground stores are substantial. Reducing these risks to acceptable levels is going to be very expensive. If the world's richest country has reservations, then the prospects for developing this technology by smaller econo-mies such as Australia are not promising. Nevertheless, undeterred by the US experience, Australian federal and state governments have committed over two billion dollars to CCS.[36]

The interdisciplinary expert study on The Future of Coal from the Massachusetts Institute of Technology envisages that coal with CCS may begin to make a noticeable contribution on a global scale around 2025 and may overtake renewable energy on a global scale around 2045.[37] CCS should be developed if possible, but meanwhile we must put more resources into implementing the safe, greenhouse friendly technologies that are already commercially available (energy efficiency, solar hot water, wind power, bioelectricity and solar elec-tricity) and rapidly develop those that are close to fruition, such as hot rock geothermal and marine power.

FALLACY 5

Nuclear power is part of the solution

The fifth fallacy states that nuclear power should be considered a viable option for cutting greenhouse gas emissions.

Response: Current reserves of *high-grade uranium ore* will only last several decades at the current usage rate. Once they are used up, low-grade ore will have to be utilised. This means that to produce 1 kilogram of yellowcake (a type of uranium concentrate), 10 tonnes or more of rock will have to be mined and milled, using fossil fuels. Under these circumstances, the CO_2 emissions from the nuclear fuel chain will be significant,[38] possibly comparable with those of an equivalent combined cycle gas-fired power station.[39]

In theory, this limitation could be overcome by shifting to fast breeder reactors, which produce so much plutonium that they could multiply the original uranium fuel by a factor of up to 50. Climate scientist James Hansen is pushing for the US to redevelop this kind of technology rapidly.[40]

The world's last large fast breeder reactor, the French Superphénix, operated only 278 days of full-power equivalent between 1986 and 1997. It was closed in 1998, after many technical problems and costing about €9 billion.[41] At present there are no commercial scale fast breeders operating. The Russian 600 MW demonstration *fast (neutron) reactor*, Beloyarsk, operates intermittently, but it too has a history of accidents and does not seem to have ever operated as a breeder.[42] The pro-nuclear MIT study does not expect that the breeder cycle will come into commercial operation during the next three decades.[43]

Even if another fast breeder power station were to be built in the future, large-scale chemical reprocessing of spent fuel would be necessary to extract the plutonium produced in its core. Since spent fuel is intensely radioactive, reprocessing has its own serious hazards and costs.[44] In the USA, three 'commercial' reprocessing plants were built at various times, but none was commercially viable and they were closed permanently. The UK reprocessing plant at Sellafield (formerly

Windscale) was shut down in April 2005, when it was discovered that
83 000 litres (equivalent to half an Olympic swimming pool) of highly
radioactive liquid had been leaking from it unnoticed for the previous nine months.[45] As of February 2009, Sellafield has been sold to a
US-based consortium and its future is unclear. Non-military reprocessing is still carried out on a large scale at La Hague in France, at a
new plant in Japan and at small plants in India and Russia. In practice,
only a small fraction of plutonium produced globally in nuclear power
stations is being 'recycled'. The rest remains unseparated in high-level
nuclear waste, stored 'temporarily' (for the past 50 years in some cases)
next to the power stations that produced it.

The latest nuclear 'solution' being promoted is the proposed
integral fast reactor, whose main selling point is that it combines the
capacity to breed plutonium with an onsite reprocessing plant. This
is the rather shaky basis for the claim made by some proponents that
all the radioactive waste produced by the reactor will be consumed
onsite. While this is true for most of the long-lived *transuranics* such as
plutonium-239, it is untrue of the medium-lived fission products such
as strontium-90 and cesium-137, which form the major part of the
waste. Nevertheless, compared with an ordinary nuclear reactor, the
wastes leaving the site of an integral fast reactor would be shorter-lived
and unsuitable for nuclear weapons.[46] On-site proliferation would still
be possible, though difficult.

Another limitation of nuclear power as a potential greenhouse solution is its long planning and construction time, especially for countries that are new entrants to this complex and dangerous technology.
In Australia, government ministers and nuclear experts have admitted
that the first nuclear power station and associated infrastructure would
take 15 years to construct (if there was no public opposition).

Even in the UK, which already generates 19 per cent of its electricity from nuclear power, governments have kept changing the type
of nuclear power station to be built, a practice that has resulted in
long construction times and hence very high costs. The most recent
nuclear power station to be built there, Sizewell B, rated at 1.2 GW,

had a capital cost of £3.7 billion, equivalent to £3000 per kilowatt in 2005 British pounds, not counting interest during construction.[47] This extreme case demonstrates the very high financial risk of building nuclear power. One of only two nuclear power stations under construction in a Western country is at Olkiluoto in Finland. Construction commenced in 2005 and by December 2008 it was at least two years behind schedule and its cost had escalated by about €1.5 billion.[48] Long development times also apply to fast breeder reactors and reprocessing plants. Such complex and dangerous technological systems cannot be rushed. The integral fast reactor does not exist. Even with a crash development program, it could not make a significant contribution to reducing greenhouse gas emissions in the USA before 2030.

Based on existing technology, nuclear power is neither a short-term nor a long-term solution to global warming. So-called Generation IV nuclear power is unlikely to be able to make a significant contribution before 2030, if ever. Nuclear power is a diversion from genuine solutions to global warming.

FALLACY 6

Spent fuel doesn't contribute to weapons

This fallacy states that the spent fuel from nuclear power stations cannot be used to make nuclear weapons.

Response: When I was an activist against nuclear weapons and nuclear power, this false claim was the one most frequently uttered by the nuclear industry and its supporters. It has been refuted by many experts, including leading US nuclear bomb designer Dr Theodore Taylor, commissioner of the US Nuclear Regulatory Commission Dr Victor Gilinsky and the US Department of Energy.[49] A standard 1000 MW nuclear power station produces about 200 kilograms of *reactor-grade plutonium* annually, enough for 20 nuclear bombs. Although the explosive yield of a nuclear bomb charged with reactor-grade plutonium is less predictable than from *weapons-grade*, it is still a very destructive weapon.

It has also been claimed incorrectly that a nuclear power station based on thorium rather than uranium cannot produce a nuclear explosive. In fact, to use thorium as a fuel, it must first be converted to uranium-233, which is *fissile* (able to undergo nuclear fission) and so can be used either to fuel a nuclear reactor or provide the explosive in a nuclear bomb.

Nuclear power and nuclear weapons are intimately linked. This has been the case in the USA, UK, France, China, India, Pakistan, North Korea and now Iran. In addition to the dual uses of nuclear materials, training engineers and technicians for nuclear power provides most of the training required to develop nuclear weapons.[50] In the words of a Greenpeace UK spokesperson, 'Reaching for nuclear power to solve climate change is like taking up smoking to lose weight.'

FALLACY 7

Efficient energy use has little potential

This fallacy states that energy efficiency cannot contribute adequately to fighting climate change. The message is: individuals, businesses and industries should simply not bother reducing their carbon footprint.

Response: This fallacy is propagated mainly by neoclassical economists with blind faith in perfect *competitive markets*. 'Top-down' *macroeconomic* models depend on this invalid assumption. As a consequence, economists who believe in these models claim that the market automatically incorporates all cost-effective improvements in energy efficiency based on existing technologies and there is nothing left to do. In reality, study after study, based on an analysis of the technologies and processes for using energy, identify the huge potential for cost-effective energy use, which has been held back by *market failure*. Examples of market failure are the split incentives between landlord and tenant; lack of information; lack of appropriate institutional structures, such as *energy service* companies; and other barriers.[51] Energy efficiency is the cheapest and fastest greenhouse gas reduction measure in the energy sector. Such efficiency will increase rapidly once governments introduce regulations and standards for energy labeling and minimum energy performance standards for *all* buildings, appliances and energy-using equipment. A ban on new conventional coal-fired power stations is also essential.

The huge potential of energy efficiency, backed by effective government policies, is demonstrated by comparing the per capita electricity consumption of California with the whole of the USA. In 1973, during the OPEC oil embargo, the national figure was 15 per cent higher than California's. By 1980, California's per capita electricity consumption had leveled off at about 7 MWh per year and remained approximately at that level until after 2000. However, the US figure continued to increase until by 2000 it was 12 MWh per year, 70 per cent higher than California's. The difference resulted from long-term policies by previous and present Californian governments, both Democrat and Republican, to encourage energy efficiency.[52]

A related fallacy, again propagated by neoclassical economists, is the *rebound effect*. The argument is that, in the usual situation where energy efficiency saves people money, they will spend all the savings on using more energy. This is the alleged 'rebound'. This argument is a gross exaggeration of the actual situation. In reality, only a small fraction of each dollar spent by consumers increases energy use. The dropped ball doesn't bounce all the way back to the hand that releases it. This fraction can be made even smaller by means of the policies mentioned above. It can be reduced to zero by providing consumers with 'packages' of energy efficiency and renewable energy such that the total net cost of each package is zero and therefore so is the rebound. Such packages could achieve big reductions in greenhouse gas emissions, as shown by the McKinsey *cost curves*, which rank the costs of various energy efficiency and energy supply measures, showing the corresponding potential reductions in greenhouse gas emissions from each measure. In scenario studies for cutting greenhouse gas emissions by 30 per cent by 2020, McKinsey & Company found that the economic savings from energy efficiency are so large that they can pay for most of the additional costs of cleaner energy supply.[53]

Renewable energy cannot provide base-load power

This fallacy states that renewables such as wind and solar can't provide *base-load* power (that is, a 24/7 supply of electricity). Furthermore, it implies that base-load is the only important type of power station.

Response: As I travel from town to town, addressing public meetings, community groups, media, professional and business seminars, trade union meetings, political party branches, and academic conferences and seminars, this is the most frequently raised misconception about renewable energy that I have to counter. Its prevalence is the result of a massive international propaganda campaign by vested interests to denigrate renewable energy, in an attempt to keep this rapidly developing set of clean technologies out of mainstream electricity production. In its most simplistic form, the fallacy is expressed (eg, by Australia's former prime minister, John Howard) as, 'the Sun doesn't shine at night and the wind doesn't blow all the time'. The refutation requires somewhat deeper understanding.

Because it is too expensive to store electricity on a large-scale, electricity generation systems have to follow the variations in demand. To do this, conventional systems have two principal types of power station: base-load and *peak-load*.[54] Base-load power stations operate 24 hours a day, seven days a week, except when they break down. They have high capital cost and low running costs. Conventional base-load power stations are generally coal or, in some countries, nuclear. They take all day to start up from cold and are inflexible in terms of following the daily variations in demand. On the other hand, peak-load power stations are expensive to operate, but are very flexible in terms of meeting the daily peaks and covering unexpected breakdowns in base-load power stations. Conventional peak-load power stations are either gas turbines (like jet engines) that burn natural gas or oil, or hydroelectric power stations.[55] To follow variations in demand, a conventional electricity-generating system needs both base-load and peak-load power stations.[56]

Contrary to the fallacy, several types of renewable energy can substitute directly for conventional base-load coal and nuclear power. Bioelectricity (electricity from the combustion of organic material such as crop residues), *solar thermal electricity* with low-cost thermal storage, and geothermal power can all take on the role of base-load. Some locations (such as, Sweden, Iceland and Tasmania in Australia) have so much hydroelectric potential that it can provide base-load too.

Energy efficiency and solar hot water can also substitute for some base-load power stations. In several countries, base-load coal-fired power stations are used to provide off-peak electric water heating from midnight to dawn, when electricity demand would otherwise be very low. This allows the inflexible base-load power stations to be operated 24/7. However, if off-peak electric hot water were terminated and replaced with solar, gas and electric heat pump hot water, several coal-fired power stations could be retired or not built and millions of tonnes of CO_2 emissions would be saved per year.[57]

Even large-scale wind power, from geographically dispersed wind farms, can be made as reliable as base-load coal or nuclear power by adding a little peak-load power capacity, which does not have to be operated frequently. The more geographic dispersion, the more the total wind power generation is smoothed out.[58]

Since gas turbines can burn renewable liquid and gaseous fuels made from *biomass* (organic materials), renewable energy can provide peak-load as well as base-load power. Furthermore, since most electric power is used during the daytime, daytime power (from *intermediate-load* and peak-load power stations) is at least as important as base-load. Even in the absence of low-cost electrical storage, solar photovoltaic (PV) electricity will be able to make a large contribution to daytime power as its price declines in the future. In addition, solar thermal power stations with low-cost thermal storage can be operated in both base-load and peak-load modes, depending on electricity prices at different times of the day (see chapter 3).

Thus, energy efficiency and renewable energy can provide both base- and peak-load. However, it must be said that these concepts of base- and peak-load were developed for conventional generating systems based on fossil fuels. Renewable energy blurs the boundaries between these concepts. The important thing is that the combination of energy efficiency and several types of renewable energy can provide a generating system that is 100 per cent renewable and just as reliable as a conventional generating system.

FALLACY 9

Renewable energy has huge land requirements

This fallacy states that renewables such as wind and solar take up a great deal of space.

Response: Actually, wind and solar power generally have smaller land requirements than equivalent coal power with open-cut coal-mines.

Wind power is normally installed on agricultural land, where its turbines and access roads occupy only 1–2 per cent of land area. The other 98–99 per cent of land can still be used for agriculture. To replace a 1000 MW coal-fired power station with wind power would require 5–20 square kilometres (km) of land, depending upon wind speeds at the wind farm sites.[59] Typical open-cut coal-mines occupy over 50 square km. Even underground coal-mines, using longwall mining technologies, sometimes damage quite large areas of land on the surface.

A square of area 108 km x 108 km = 11 700 square km could supply from solar energy all the USA's 2008 electricity generation. This area corresponds to one-eighth of 1 per cent of US land area. A square of area 22.6 km x 22.6 km = 510 square km or 0.66 per cent of Australia's land area could supply from solar energy all of that country's 2005 electricity demand. My back-of-the-envelope calculations assume a conversion efficiency of 20 per cent and no solar concentrators. With concentrators, even smaller areas would be required. In Australia the residential component of electricity demand could be supplied by covering about 28 square metres (m) (5.3 m x 5.3 m) of rooftop space of each house with flat-plate solar PV modules. With energy efficiency, this area could be halved. Thus, no additional land would be required for residential solar electricity and the land required for commercial and industrial uses of electricity would be modest.

In practice, neither wind nor solar would supply all electricity, which would be provided by a broad mix of renewable sources.

Sustainable energy is more costly than fossil fuel energy with CCS

The fallacy states that a sustainable electricity-generating system is more expensive than a fossil fuel system with coal power with CCS as base-load.

Response: As the ExternE studies[60] and the Stern Report[61] recognise, conventional coal power is very expensive in terms of economic, environmental and health impacts. The costs of drought, increasing prevalence and severity of bushfires, loss of tourism at snow-fields and coral reefs, and the impacts of rising sea-levels on urban infrastructure will be huge. At present these costs are not included in the price of coal power in the USA or Australia. They are externalised, namely imposed on the environment and society. Carbon pricing, by means of a *carbon tax* or emissions trading, is a means of internalising at least some of these *external costs*, as would passing a law to mandate that all new coal-fired power stations must have CCS.

Before a carbon price is implemented, all clean supply-side alternatives appear to be more expensive than dirty coal power. However, the combination of efficient energy use and renewable energy is going to be much less expensive than coal with CCS without energy efficiency. As discussed under fallacy 7, the economic savings from efficient energy use can compensate for much of the additional costs of renewable energy. Another way of expressing this is that, although the cost of a kilowatt-hour of electricity will increase, the number of kilowatt-hours used will decline and so the total energy bill will not necessarily increase significantly. This result is illustrated by several studies: for example, McKinsey & Company found that, with a combination of demand-side and supply-side measures, Australia could cut its greenhouse gas emissions to 30 per cent below the 1990 level by 2020. The average annual gross cost would be A$290 per household, while the expected increase in annual household income over the period would be A$20 000.[62]

The estimated cost of $290 could also be compared with the hypothetical case of CCS costing $60 per tonne of CO_2 in 2020. (Both the assumptions of existence of commercial CCS in 2020 and its low cost are generous to CCS.) Then, assuming an annual average household electricity demand of 8 MWh, with 1 MWh emitting 1 tonne of CO_2, leads to an additional annual cost of $480. On this basis, energy efficiency plus renewable energy is clearly cheaper than CCS without energy efficiency.

If proponents of CCS claim that they too can obtain the benefits of energy efficiency, you can point out that energy efficiency has not been implemented to a significant degree with coal power. Indeed, one purpose of developing coal with CCS is to maintain endless growth in electricity demand. Under these circumstances, it is unlikely that more than lip service will be paid to energy efficiency, which is the present situation outside Europe.

FALLACY 11

Cutting emissions = economic collapse

This fallacy states that making deep cuts in greenhouse gas emissions would cause economic collapse. In the words of Mitch Hooke, chief executive of Minerals Council of Australia: 'That is the equivalent … of moving to a candles' economy, riding horses. You've got to shut down your transport sector and your power generation … that's the magnitude of the challenge.'[63]

Response: I have refuted this at the microeconomic level (within the energy sector) in the response to fallacy 10. I now address the issue at the macroeconomic (global and national) level.

A major global assessment was conducted by Sir Nicholas Stern, the former chief economist of the World Bank. The Stern Review found that the total cost of climate change, resulting from a *business-as-usual* scenario, would be equivalent to a reduction of about 20 per cent in per capita consumption, now and into the future. The review describes this economic impact as being 'on a scale similar to those associated with the great wars and the economic depression of the first half of the 20th century'. In contrast, Stern found that the cost of stabilising greenhouse gas concentrations in the atmosphere at 500–550 ppm of CO_2-equivalent to be about 1 percent of annual global GDP by 2050.[64]

Although several climatologists now consider a safer target to be in the range 300–350 ppm, the results of the Stern Report are encouraging for climate action, especially when we consider the other economic benefits from cutting greenhouse gas emissions that were not considered by Stern. These result from:

- reduced air pollution and associated respiratory diseases;
- reduced water consumption and land degradation by coal-fired power stations and coal-mines;
- reduced prevalence of road crashes and congestion by increased use of urban public transport;
- better health resulting from more walking and cycling; and
- increased energy security.

In Australia, Treasury conducted extensive macroeconomic modeling of the economic consequences of Australia reducing its greenhouse gas emissions according to various scenarios. In particular, Treasury found that a 25 per cent cut in emissions would reduce the annual per capita gross national product growth over 2010–50 by only 0.2 per cent compared with an optimistic reference scenario. In relation to targets of up to 30 per cent reductions by 2020, Treasury and several other studies found GDP reductions in the range 1–2 per cent below the reference scenarios by 2020.[65] This is hardly economic collapse!

Most macroeconomic models, including those used by Treasury, assume the existence of a competitive market. But, as mentioned above, energy efficiency suffers from several market failures, which violate this assumption. As a result, many cost-effective energy efficiency measures are undervalued or omitted by these models. This means that the models tend to overestimate the costs of greenhouse response.[66]

Unlike Stern, none of the Australian macroeconomic studies attempted to calculate the costs of business-as-usual. If Stern is correct, the costs of deep cuts in emissions could be much less than the costs of business-as-usual.

Of course, if climate change accelerates, very rapid reductions in emissions will be needed, with the result that many existing fossil-fuelled power stations and some large greenhouse-intensive industries will have to be retired well before the ends of their operating lives, becoming *stranded assets*. This situation could greatly increase the costs of mitigation, but then, the costs of doing nothing would also be much greater.[67]

FALLACY 12

Energy efficiency and renewable energy will cost jobs

This fallacy states that substituting energy efficiency and renewable energy for coal-fired electricity would lose jobs.

Response: Like fallacy 8, this falsehood has been disseminated by a massive propaganda campaign by fossil fuel interests. The hypocrisy of industries, that previously encouraged governments to fire vast numbers of workers in electricity industries around the world in the name of corporatisation, privatisation and economic efficiency, is breathtaking. Now they suddenly claim to be concerned about jobs!

The truth is that energy efficiency and renewable energy can provide several times more *local* jobs, per kilowatt-hour of electricity generated or saved, than coal. This is because the smaller scale of sustainable energy technologies (compared with coal) lends itself to local manufacture.

For example, when a wind farm is built in Australia, over 50 per cent of the capital cost is spent in Australia. As the wind industry grows (for example, under the stimulus of an expanded Renewable Energy Target), the Australian content could grow to 75 per cent. Wind power currently employs in Australia 2–3 times the number of job-years per kilowatt-hour of coal power (including the associated coal mining), while bioelectricity employs 3.5 times (mostly in rural areas).[68] Energy efficiency technologies and measures also employ several times more job-years.[69]

When a coal-fired power station is built in Australia, only about 25 per cent of the capital cost is actually spent in Australia. Similarly, large coal-mining equipment, such as dredges for open-cut mining and longwall diggers for underground mining, is imported. As the result of automation, employment in coal mining in Australia has nearly halved since 1986, even though the quantity of coal mined has increased substantially.

It is simple to show that the job losses from the Australian coal industry from a 25 per cent renewable energy target could be addressed by not replacing a small fraction of the workers who retire annually from the coal industry. According to data from the Australian Bureau of Statistics, this industry currently employs directly about 30 000 people in Australia. Taking account of the fact that 80 per cent of Australia's coal is exported, there are only about 6000 workers employed in coal mining for coal use in Australia. If renewable energy is increased from its 2008 level of 8.5 per cent to 20 per cent of Australia's electricity by 2020 (as in the proposed expanded Renewable Energy Target) and if it all substitutes for coal power, this means that 11.5 per cent of 6000 direct coal jobs or 690 jobs would be affected. Over the 12 years from 2009–20 inclusive, this is 58 coal jobs per year, less than one-tenth of the expected annual retirements from the coal industry.

Even allowing for a generous multiplier factor of four for indirect coal employment would not change the qualitative result: job losses in the coal industry are easily accommodated by retirements and many more jobs will be created in renewable energy. In practice, as coal-mines and coal-fired power stations are closed, jobs would be lost in steps. Therefore, only transitional arrangements would be required from federal and state governments to assist workers, who have been prematurely retired, with retraining, relocation and pensions.

A carbon price nullifies need for other policy measures

This fallacy claims that, once a country has a carbon price, implemented either by a carbon tax or a *cap-and-trade* emissions trading scheme, other policy measures to assist efficient energy use and renewable energy are unnecessary. We should let the market determine the choice of technologies.

Response: This is an example of the narrowest type of neoliberal economic dogma, uninfluenced by reality. It should be rejected firmly by climate action campaigners. The fallacy assumes that the energy sector is a perfectly competitive market and so price determines everything. In practice, efficient energy use is characterised by several market failures, as pointed out in the response to fallacy 7. Even efficient energy use measures that are highly cost-effective are not being widely implemented in the absence of regulations and standards for buildings, appliances and energy-using equipment.

The market also fails to provide the necessary infrastructure for renewable energy, rail transport and public transport. Fundamentalist economists fondly imagine that research institutes, *transmission lines*, railways and gas pipelines somehow arise from the primeval ooze, as long as the price is right. This rarely happens in practice. Government funding is needed for both the hardware and software, which includes good design and planning.

The market, by its nature, can't provide long-term planning. A long-term greenhouse target for industrialised countries, consistent with greenhouse science, is an 80–95 per cent reduction in emissions by 2050, with such a level maintained for centuries. At present the only commercially available technologies that can do this in the energy sector are energy efficiency and renewable energy. This means that, to be sure of reaching the 2050 greenhouse target, we must plan on the basis of obtaining a major contribution from energy efficiency and renewable energy as soon as possible. This will be difficult in the face of the present determined lobbying of governments by major greenhouse gas polluters. These groups demand a weak cap on emissions and hence a low carbon price in emissions trading schemes.

A low initial carbon price of (say) $25 per tonne of CO_2 may just be sufficient to allow natural gas in some countries to compete with dirty coal for base-load electricity. However, such a low carbon price would not enable renewable energy to compete with coal or gas. In effect, without complementary measures to a carbon price, renewable energy could be held back for decades until gas becomes scarce and its price very high, or until the carbon price gradually becomes high. In the short term, natural gas can play a valuable role as a back-up to solar hot water, solar industrial heat, solar thermal electricity and wind power. But, if it is encouraged to take on the principal energy task, it will block the dissemination of renewable energy.

Another way of looking at this is to recognise that a well-designed emissions trading scheme fosters the economic optimal mix[70] of technologies at a particular carbon price, which means at a particular time. But, we have to plan for a variety of times, such as 2020, 2030 and 2050, which will in general have higher carbon prices and hence different technology mixes. Therefore, it is vital that policy measures for greenhouse gas mitigation promote the development *now* of a range of zero emission technologies with high potential. Although solar electricity is expensive now, it has huge potential for the longer term. Since it will benefit little from an emissions trading scheme for the foreseeable future, because the price gap is too large, it needs separate policy measures until its costs are much lower. Solar power is the least likely of all renewable energy sources to benefit from carbon pricing. It needs both further research and development and incentives for expanding the market. The latter goal can be fostered by means of *gross feed-in tariffs* (see chapter 4).

Another weakness of the market is that it doesn't handle risk well. The energy technologies available in our portfolio carry different degrees of risk. We know that energy efficiency and most kinds of renewable energy work and are very safe. We know the current costs of several renewable energy technologies and can make reasonable projections of future cost reductions as the size of their market increases. There is much more uncertainty about the performance, safety and future costs of CCS and Generation IV nuclear power. In practice, several countries are already funding CCS heavily, while giving only modest support to renewable energy. Examples include the USA under the former president, George W Bush (2001–09), and Australia under the former prime minister, John Howard (1996–2007), and Prime Minister Kevin Rudd during his first year in office (2007–08). Renewable energy, as a set of low-risk technologies that work and are safe, should at very least receive matching support to CCS.

Finally, the possibility must be considered that the form of carbon pricing most favoured by industry and government, cap-and-trade emissions trading, may fail to be effective in reducing emissions. Therefore, in the words of my electrical engineer colleague at UNSW, Iain MacGill, we need 'policy insurance'. This can be provided by so-called 'complementary measures', such as renewable energy portfolio standards, tax deductions and gross feed-in tariffs (see chapter 4).

FALLACY 14

Cap-and-trade will cut emissions

This fallacy states that a cap-and-trade emissions trading scheme is the certain method of cutting emissions.

Response: The trading nations of the world have very little experience with emission trading schemes. We can only draw upon the US scheme for reducing sulfur dioxide (SO_2) emissions from power stations (as part of its program to reduce acid rain), and the first few years of the European Union's emission trading scheme for CO_2 (2005–).[71]

The first phase of the EU's emission trading scheme demonstrated how vested interests gained concessions that greatly weakened the scheme: notably over-allocation of emission permits in several countries and free emission permits for all.[72]

The Australian emission trading scheme, misnamed the Carbon Pollution Reduction Scheme, is due to commence in mid-2011, provided it is passed by parliament.[73] Its unconditional greenhouse target[74] is a 5 per cent reduction in emissions below the 2000 level. According to economic modeling by Treasury, this very weak target is unlikely to produce an initial carbon price of more than A$25 per tonne of CO_2. As mentioned in the response to fallacy 13, this price is too low to drive significant change in the energy supply system. At best, some black coal will substitute for some brown coal and a little more gas will be used for electricity generation and heating, but even this is unlikely because the long-term prognosis is that gas prices will rise steeply as more Australian gas is traded on the world market. The cap on the initial carbon price of A$40 per tonne of CO_2 will exclude renewable energy for several years, unless it receives support from the expanded Renewable Energy Target or other policy mechanisms. As discussed in the section on vested interests (above), Australia's biggest greenhouse polluters will receive free emission permits.[75]

These huge concessions to the principal polluters are likely to reinforce the existing sources of greenhouse gas emissions. Under these circumstances, it is difficult to see how even the 5 per cent greenhouse target will be reached. Perhaps the government imagines that most emission reductions will be achieved in the residential sector, from energy efficiency and solar hot water. Although these technologies are already cost-effective in most of Australia, they are inhibited by market failures. They cannot spread rapidly in the community unless governments enact regulations and standards, together with incentives to landlords. Renewable electricity technologies will need support from renewable energy portfolio standards (of which Australia's Mandatory Renewable Energy Target is an example), gross feed-in tariffs, tax concessions, and government funding for research, development, demonstration and infrastructure. These policy instruments are discussed in chapter 4.

The three conclusions concerning fallacy 14 are:

- Carbon pricing is necessary but not sufficient.

- An emissions trading scheme may be the wrong method of applying a carbon price, since it can easily be made ineffective under pressure from vested interests.

- The first (and possibly the second) phase of the EU emissions trading scheme were at best only slightly effective in reducing emissions, while the proposed Australian scheme is likely to be ineffective for at least several years after it commences in 2011.

5 per cent target = 25 per cent in per capita terms

This fallacy states that the Australian government's unconditional greenhouse target of a 5 per cent reduction by 2020 may appear to be small, but it is equivalent to a much larger per capita reduction of 25 per cent, when projected population growth is taken into account. In the process of Contraction and Convergence, it is the per capita emissions that count.

Response: This fallacious statement has been made several times by Ross Garnaut, author of the Garnaut Climate Change Review.[76] He is attempting to justify the very low unconditional greenhouse target that he recommends for Australia: specifically a 5 per cent reduction in emissions by 2020 compared with the 2000 level. Following Professor Garnaut's recommendation, this low target was adopted by the Australian government. Garnaut's statement involves a misunderstanding of the concept of Contraction and Convergence, in which the industrialised countries reduce their per capita emissions and the less developed countries increase their per capita emissions until each country converges to the same safe level of average per capita emissions at some future date, such as 2050.[77]

The problem with this statement is that it ignores a key point in the concept of Contraction and Convergence, namely that the population level used to calculate per capita emissions of each country must be chosen for a year *before* the beginning of the process. Otherwise Contraction and Convergence will give countries a perverse incentive to increase their populations, which will in turn increase their greenhouse gas emissions. In calculating the per capita reduction for 2020, Garnaut ignores this fundamental point and projects Australia's population to 2020 by assuming the present very high population growth rate. In statements to the media, he has supported the policy of continuing a high immigration rate for Australia, a policy that would be disastrous if applied to developing countries.

Actually, high population growth is a particularly inappropriate policy for Australia, the country with the highest per capita emissions in the industrialised world at about 28 tonnes of CO_2-equivalent emissions per year. Every additional Australian on average has a bigger greenhouse impact than an additional person almost anywhere else in the world. Therefore, until Australia's per capita emissions have contracted to an agreed safe global level, which is around 2–3 tonnes per person, it is essential that Australia implements policies to stabilise its population as soon as possible.

FALLACY 16

Lots of small changes = large emission reductions

This fallacy states that the best pathway to a large emission reduction target is through lots of small changes to the existing energy supply system.

Response: Small improvements to existing, greenhouse-intensive, energy supply systems are a waste of time and money. The reality is that these systems have to be phased out as rapidly as possible. An example is the drying of brown coal, so that its emissions are reduced (compared to those of black coal when combusted in a conventional power station). This pointless approach is popular with the Victorian government (in Australia), where brown coal dominates. Another example is making improvements to the boiler of a black coal-fired power station that lifts its *thermal efficiency* from 36 to 40 per cent. It is time to recognise that there is no future for any kind of coal-fired power station that does not capture and sequester CO_2. It would be far better to spend the available resources on changing the whole energy system to one based on the efficient use of renewable energy, than fiddling around the edges.

Examples of reducing energy demand are given in the book *Natural Capitalism*.[78] The authors show that it is often far more effective and cost-effective to design energy efficiency into new buildings, appliances and equipment at the start, than to try to modify poorly-designed existing systems. Having said that, we still face a huge retro-fitting task. Even so, some retrofitting examples where a big step is required, such as constructing a light rail system on existing major roads or banning the sale of electric resistance hot water systems, are more likely to be a more cost-effective way of cutting emissions, compared to taking many little steps.

The inadequacy of setting a small greenhouse target in a cap-and-trade emissions trading scheme is pointed out in the response to fallacy 14.

FALLACY 17

If high polluters can't pollute here cheaply, they'll move elsewhere

This fallacy is common in Australia and Europe. It states that, if *emissions-intensive trade-exposed (EITE) industries* are not given free emissions permits, they will close down their operations in Australia/ Europe and move to a less developed country. The environmental standards there will be lower and greenhouse gas emissions will be higher.

Response: This phenomenon is known as *carbon leakage*. Both the probability of it occurring and its consequences are exaggerated by those who invoke it. In Australia, the principal industries concerned are smelting of aluminium and other metals, alumina, steel, liquefied natural gas production and gold. In general they have very large plants, costing hundreds of millions to billions of dollars. They are very unlikely to uproot these plants at great expense from a politically stable country to move to a less politically stable country, where living conditions may be less attractive to management and professional staff. Furthermore, in Australia, aluminium smelting is already heavily subsidised. In most states the magnitude of the subsidies is confidential. However, in Victoria, an Auditor-General's report estimated the subsidy to the Alcoa smelter in terms of cheap electricity was worth about A$200 million per year. In addition, the Victorian government built a transmission line from the coal-fields in the east across the state to feed power to the smelter at Portland in the west and permitted Alcoa to use it free of charge. The Queensland government sold a coal-fired power station to the Comalco smelter at Gladstone for a price that energy experts believe was about half its true market value.[79]

In some cases it may be good for both the environment and the economy if the energy-intensive industries leave, because Australia is the only country where most aluminium smelting is done with coal-fired electricity. In most other countries, hydroelectricity, usually with much lower greenhouse gas emissions, is used.

Conclusion

Now that these fallacies have been exposed and refuted, it is clear that technological solutions to the climate crisis are available for the energy sector. Indeed, the only available technological solutions with negligible greenhouse gas emissions are efficient energy use and renewable energy. They are clean, safe and effective. Claims that they cannot be used as the basis for a modern industrial state are simply propaganda that is designed by vested interests to mislead decision-makers and the public. Campaigners for climate action can confidently promote these solutions, drawing upon the above refutations of the fallacies.

The technologies favoured by vested interests and their supporters in government – nuclear power and coal power with CO_2 capture and sequestration – are immature,[80] dirty and dangerous. They cannot make a noticeable difference to the climate crisis before 2020. Yet the power-holders in many countries are trying to delay the dissemination of energy efficiency and renewable energy until their favoured technologies are ready. They are giving only token support to the genuine solutions and are implementing types of emissions trading schemes that will be ineffective for several years at least. This is a recipe for disaster. Now is the time for emergency programs to implement energy efficiency and renewable energy, as discussed in the next chapter.

3

TECHNOLOGIES FOR STOPPING GLOBAL WARMING

A radical idea, whose time has come, is to phase out the technologies dominating and damaging nature and people – those based on coal, oil, gas and uranium – and phase in the technologies that work with nature – those based on sunshine, wind, biomass, wave, tide and Earth's heat. The good news, that climate action campaigners can spread to the community, is that by using energy more efficiently and better harnessing renewable energy, backed up temporarily with natural gas, we can make substantial reductions in greenhouse gas emissions. And we can do it before 2020. Technological improvements can transform energy and transport, and may soon be able to reduce greatly the emissions from agriculture too. They can buy us time, so that we can come to grips with the other, more difficult, driving forces of greenhouse gas emissions: growth in economic activity and population.

All solutions – that is disseminating greenhouse friendly technologies, changing the economic structure of a nation and controlling population growth – need new government policies for implementation (see chapter 4). Getting the right policies from government needs unrelenting pressure from the climate action movement, all of us, as discussed in chapters 5 and 6.

Solutions for energy

This section summarises some key features of the principal technologies that have been proposed for cutting greenhouse gas emissions in the energy sector. For more detail on each technology's potential resource, status of development, future development potential, economics, and environmental, health and social impacts, see my previous book, *Greenhouse Solutions with Sustainable Energy*[1] and the textbooks by Godfrey Boyle[2] and Bent Sørensen.[3] I start with the lowest cost technologies and proceed to higher-cost solutions. Here I'm using 'technologies' in a broad sense to cover hardware, software, design and associated processes.

Energy efficiency and conservation

The myriad technologies for using energy less wastefully, which can be implemented at least cost and greatest speed, fall into the category of efficient energy use (shortened to energy efficiency). Energy efficiency is having the same energy services (such as a warm house in winter or cold food) while using less energy. It is achieved by changing hardware and processes that use energy. For example, in residential buildings energy efficiency technologies include good design and choice of building materials, insulation, draught exclusion in winter, draught creation in summer, eaves to exclude summer sunshine, windows located to catch winter sunshine, thermal mass (in non-tropical regions), water efficient taps and shower-heads, ceiling fans and evaporative coolers instead of air conditioners, and efficient lighting and appliances.

Most energy efficiency technologies and measures on the market save money over their lifetimes, in the sense that their reduced energy costs pay for their increased capital costs and more. They offer cost-effective potential to cut energy consumption per capita by 25–50 per cent, depending upon location. However, market failure is endemic and so regulations and standards are essential for achieving the full economic potential across a state or nation.

Strictly speaking, *energy conservation* is not the same as energy efficiency. Energy conservation involves reducing the demand for energy services. It requires behavioural change, for example, buying fewer consumer products, wearing warm clothes instead of heating your home excessively in winter, taking shorter showers and turning off lights. To implement it on a large (eg, national) scale is more difficult than implementing energy efficiency, because energy conservation requires a cultural change, returning to the practice of thrift.

Solar hot water, solar space heating and fuel substitution

In areas where off-peak electricity is used to heat water, solar hot water, preferably gas-boosted, can achieve significant reductions in emissions and replace coal-fired power stations.

Fuel substitution can also play a role. It is generally far more efficient in energy and greenhouse terms to provide heat for a building or an industrial process directly by burning gas on-site than by using coal-fired electricity from the *grid*. It is even better to have *cogeneration* of heat and electricity on-site, or *trigeneration* of heating, cooling and electricity on-site. Solar cogeneration and trigeneration systems are currently under development: for example, a demonstration solar PV concentrator system is supplying both electricity and hot water to the Bruce Hall, one of the student residences at the Australian National University in Canberra.[4] For solar space heating alone, the economics are generally unfavourable, unless the system can provide year-round benefits by running summer cooling as well.

Wind power

Averaged over the three decades commencing 1978, wind power has been the fastest growing electricity generation technology in the world. At the end of 2008, global installed generating capacity was 120.8 GW, that is, 120 800 MW. In 2008 wind power grew by 28.8 per cent and, over the 30-year period from 1978, the annual growth rate was consistently in the 25–30 per cent range. Regions with high wind energy potential include northern Europe, the Midwest and Rocky Mountain states of the USA, north and north-west China, and southern Australia. The countries with the largest installed wind power capacities are the USA (which overtook Germany in 2008), Germany, Spain, China and India in that order. Australia ranked 13th with 1306 MW. In terms of economic value, the global wind market in 2008 was worth about €36.5 billion in new generating equipment.[5]

As the least expensive of the new, renewable sources of electricity, wind power is the backbone of the EU's renewable energy target and in 2008 was the leading power source for new generation capacity there. Given a suitable wind regime and support from government policies, many countries, including the USA,[6] could generate at least 20 per cent of electricity from the wind by 2030. Denmark already achieved that wind penetration in 2003.

The wind power industry's great success has attracted opposition by vested interests, especially the coal and nuclear industries, which circulate anti-wind misinformation mainly through front groups. The most frequently repeated fallacy is that wind power cannot substitute for base-load power stations such as coal and nuclear.[7] The anti-wind material is further disseminated by NIMBY (Not In My Back Yard) groups, many of which purport to be environmental groups, although very few of their leaders are recognised to be genuine environmental activists. In reality, no energy source is perfectly benign, but wind has one of the lowest measurable environmental impacts of all sources of electricity.

At 20 per cent wind energy penetration into the grid, a modest amount of back-up may be required from peak-load power stations, either hydroelectric or gas. The greater the geographic spread of wind farms, the less back-up is required. The back-up power stations are not run continuously – rather they are brought on-line when required – and so do not add greatly to costs or greenhouse gas emissions. As wind energy penetration increases beyond 20 per cent electricity generation, the amount of back-up required increases. Denmark's back-up is its grid interconnections to Norway and Germany. In practice, it makes little difference whether back-up is provided by dedicated peak-load power stations or grid connections to the outside world. The only advantage of grid back-up is that it works in both directions, taking excess wind power away to sell elsewhere as well as supplying shortfalls in wind energy.

At appropriate sites, on-shore wind power is generally less expensive than nuclear power or impartial projections for coal power with carbon capture and storage (CCS).[8] As its market grew, the cost of wind power declined steadily until 2006. In the period 2006–08, wind turbine prices increased temporarily to a higher level, because supply could not keep up with demand in the global market. In the near future, we can expect a continuation of declining prices for wind turbines as supply bottlenecks are overcome and the market continues to grow.

Off-shore wind power is still under development and demonstration. As a result, its capital costs are at least 50 per cent higher and its operating costs 100 per cent higher than those of on-shore wind farms. In the longer term, the higher wind speeds and more consistent winds off-shore may compensate for the higher capital and maintenance costs of off-shore wind farms. From a European perspective, off-shore wind is also worth pursuing because good on-shore sites are becoming scarce and large windy areas with shallow waters (less than 40 metres in depth) are available off-shore. Existing off-shore wind turbines have their foundations in the ocean bottom. However, research and development has begun on floating tethered off-shore wind turbines, drawing upon experience with oil platforms. If the engineering and economics turn out to be favourable, then wind farms could be built in deeper waters.

Bioenergy

Bioenergy is the immediate source of metabolic energy of almost all living things on Earth. People have used it in limited ways for millennia (burning firewood for cooking and heating), for centuries (brewing alcohol from crops) and for decades (farmers extracting oil from oilseed plants to fuel their tractors). Today we have a much wider range of bioenergy options.

We define bioenergy to be useful energy produced rapidly from biomass. Biomass is material produced by photosynthesis, in which growing plants capture CO_2 from the atmosphere and water from the soil to form carbohydrates and oxygen. Biomass includes forestry and agricultural residues, dedicated crops, food processing wastes, weeds, oil-bearing plants, animal manure, and the organic fraction of municipal solid wastes. The solar energy stored in the carbohydrates may be released by burning the biomass directly, or first converting it into even more useful forms of stored bioenergy, such as the liquid fuels ethanol, methanol and *biodiesel*, or biogas (which is mostly methane), before burning.[9] The combustion of biomass can be carbon neutral (that is, produce no net greenhouse gas emissions) over a year or so, provided its rate of use is no greater than its rate of production and the conversion processes do not use significant quantities of fossil fuels.[10]

Fossil fuels, on the other hand, take millions of years to form. When combusted, they release CO_2 at a rate that is much, much faster than their rate of formation. Their use on a large scale is completely upsetting the carbon balance of planet Earth.

Biofuels can substitute for fossil fuels in a wide variety of services. Solid, liquid or gaseous biofuels may be burnt in power stations to generate electricity and heat. Liquid and gaseous biofuels, such as ethanol and biodiesel, can substitute for some of the oil used in transportation. Biomass can also provide a raw material for manufacturing fertilisers, plastics, paints and other chemicals. When heated in an atmosphere with reduced oxygen in a process known as *pyrolysis*, biomass produces *biochar*, an organic charcoal that greatly enhances the ability of soil to sequester CO_2. In all these services, biomass can reduce or sequester greenhouse gas emissions.

The potential for using liquid or gaseous biofuels for transportation is discussed in the next section. Here I focus on *stationary energy*. The residues of existing plantation forests and agricultural crops can be combusted to produce electricity and useful heat, without any additional use of land. The ash produced can be backloaded on the trucks that collect the biomass, to return nutrients to the soil. Several countries have large bioenergy potential: especially Scandinavia, Austria and Australia.

In Australia, 30 per cent of electricity could be generated from the residues of wheat, sugar cane, plantation forests and other biomass sources by 2040.[11] However, climate change may cut this potential in half. By 2020, bioelectricity could contribute 5–8 per cent of electricity, given government incentives.

Solar thermal electric power

Also known as *concentrated solar thermal power*, solar thermal electric power is now growing rapidly from tiny bases in Spain, Portugal and the USA. As more experience is gained, big cost reductions are being claimed, but it is still too early to predict accurately its cost of electricity under future large-scale mass production.

Sunlight is concentrated and focused with mirrors or lenses to heat a fluid, which produces steam to turn a turbine or run a steam engine, and so generate electricity. There are several types of concentrators, including parabolic troughs, paraboloidal dishes, fields of flat mirrors, and the Compact Linear Fresnel Reflector.[12]

A limitation of concentrator systems is that they require direct sunlight. This means that regions with diffuse sunlight (eg, humid tropical and many coastal areas) are less suitable than dry inland areas. One proposal for a very large-scale use of thermal power is to install many solar power stations in North Africa and feed the power by high-voltage direct-current transmission lines under the Mediterranean into the European grid. In countries such as the USA, Australia and northwest China, the possibility of feeding solar power from desert regions to the cities exists. However, if land is available, it may be less expensive to site the solar power stations near the cities, trading off the reduced solar power against the cost of long-distance transmission.

The great advantage of solar thermal electric power is that heat can be stored for hours or even overnight relatively cheaply in water, molten salt, graphite and by splitting ammonia into nitrogen and hydrogen. This means that solar electricity can be generated on demand, 24 hours a day, when required. It can be supplied as either base-load or peak-load power. Alternatively, the 'waste' heat can be used for heating buildings, desalination of seawater, sewerage treatment, horticulture and aquaculture.

A variety of solar thermal electric power stations and solar thermal storage are being developed and constructed. Several years of testing, experience and improvement will still be needed to determine which of these are efficient reliable technologies and which are 'lemons'.

In countries with lots of sunshine and specific policy incentives, 3–5 per cent of electricity could be generated from this source by 2020 and much more by 2030.

Solar photovoltaic power

We are all familiar with solar photovoltaic (PV) systems. We see them on roofs, powering portable road signs, and on experimental cars racing across the desert. In most countries they are still in niche markets, although they have huge potential.

PV systems have special surfaces that emit electrons when exposed to light. The moving electrons form an electric current. PV collectors may be flat modules, which are convenient for residential rooftops, walls and eaves, or they may have mirrors to concentrate the sunlight onto the special surfaces in solar power stations.

As a result of government policies to stimulate the market for PV systems in Europe and Japan, global PV capacity has been growing at about 40 per cent per year, a faster rate than wind power, since 2000.[13]

The advantage of PV is that it has no moving parts or fluids, apart from any concentrators that track the Sun. Another advantage is that it can be easily installed on residential roofs. A possible disadvantage is that it produces electricity directly and electricity is still very expensive to store on a large scale. Therefore, PV systems are a source of daytime power; they can contribute to peak-load and intermediate-load power, but not base-load. This is not a severe limitation in an integrated generating system, containing a mix of renewable electricity sources, keeping in mind that electricity demand in daytime is generally much greater than at night. Even without dedicated electrical storage, it may be feasible to generate 10–15 per cent of a country's electricity from PV. PV power would be substituting mainly for gas-fired electricity. If low-cost electrical storage becomes available (for example, in flow batteries such as vanadium redox), then PV too could substitute for base-load coal.

By 2020, if not before, PV power without storage will probably be competitive with retail electricity prices in the residential sector in many countries. This can be said with confidence for three reasons:

- the scientific and technological advances that are occurring in PV cells;

- the rapid growth in the market for solar modules; and
- the increasing prices of grid electricity from conventional sources in many parts of the world.

Additional price increases of grid electricity will result from carbon taxes and emissions trading schemes. The crossover points may be reached by 2015 in much of Europe, where grid electricity prices are high, and several years later in the USA and Australia. By 2020, 3–5 per cent of electricity could be generated from PV in sunny countries, given strong incentives from government, and much more by 2030.

Efficiencies of solar cells are increasing rapidly in the laboratory. The world record efficiency for a conventional silicon solar cell is approximately 25 per cent. Higher efficiencies (over 40 per cent) are being obtained by stacking different layers of photovoltaic materials sensitive to different wavelengths of sunlight on top of one another. Another important technological advance is the development of solar cells based on thin films of silicon. Since high-grade silicon is expensive, the ability to manufacture solar cells with one-tenth the silicon of conventional cells gives savings in both costs and energy inputs. The new low-silicon cells include the crystalline silicon on glass (CSG) cells, developed at the University of New South Wales and now being manufactured in Germany,[14] and Sliver cells, developed at the Australian National University,[15] but not yet in mass-production.

Geothermal power and heat

In volcanic regions of Iceland, the Philippines, New Zealand and California, steam is vented from the Earth's surface and used to generate electricity. This, the conventional means of generating geothermal power, is geographically limited. However, the heat of non-volcanic regions of the Earth may harness much greater potential. The new approach – hot rock geothermal power, sometimes called 'engineered geothermal systems' – is being developed in Australia, France and elsewhere.

This involves drilling 3–5 kilometres down to regions of granite that have been heated to 200–350 degrees Celsius over millions of years by the radioactive decay of traces of uranium and other radioactive elements in the rock. At least two wells are drilled close to each other and the rock between the bottoms of the wells is fractured. Once the connection is established, water is pumped down one well and superheated steam comes up the other. The steam turns a turbine to generate electricity. All of the water, now cooler, is reinjected into the first well, thus completing a closed cycle. In practice, it may be difficult to obtain the optimal flow rate to give an economic generation of electricity.

At the time of writing, two small demonstration hot rock geothermal power stations are close to commencing operation: one is the joint European project at Soultz-sous-Forêts in France;[16] the other is run by Geodynamics Ltd, in South Australia.[17] If successful, the initial small systems of 6 MW and 1 MW electric respectively will each be expanded to several tens of megawatts and then to commercial systems of several hundreds of megawatts. This scaling-up process will take time, so we cannot expect large contributions before 2020. However, hot rock technology has huge potential in Europe, North America and Australia. A study by an interdisciplinary team at the Massachusetts Institute of Technology has estimated that the USA has geothermal potential for at least 2000 times its 2005 consumption of *primary energy*, not just electricity. The MIT study proposes that, by 2050, 100 GW of electrical capacity could be developed in the USA.[18]

For generating electricity economically, rocks with temperatures of 200 degrees Celsius or more are required. However, much lower temperatures at shallow depths (tens of metres) can be used for heating and cooling buildings. Some commercial geothermal heating systems are on the market for large buildings. The potential for further development and dissemination of the technology is huge.

All forms of geothermal power and heat use are very low in environmental impacts. For hot rock geothermal power, the principal impact is likely to be from the energy input to drilling. Although geothermal power is not, strictly speaking, renewable on timescales of human interest, it is classified together with renewables because of its low impact and because each pair of wells is expected to last for two to three decades and then would recharge from the surrounding rock over a period of about 50 years.

Hydroelectric power

For millennia the power of rivers was used to grind grain and saw wood. In the 19th century hydro power became the first renewable source of electricity. Nowadays hydro provides about 20 per cent of global electricity and it is still the principal renewable source of electricity. The potential for additional hydroelectric power from large dams is constrained by the environmental and social impacts of flooding valleys. These impacts include the destruction of biodiversity, methane emitted by flooded vegetation, and the displacement of populations from their land.[19] Nevertheless, there is still much hydroelectric development occurring in China and a few other countries.

Marine power

Marine power is electric power generated from movements of the sea. These sources are less developed than wind and bioenergy. The only commercially available technology is conventional tidal power, based on building large dams across estuaries, but this is limited to a few geographical locations in the world. Just like large dams on rivers, tidal power has potentially huge environmental impacts.

However, tapping the power of tidal currents, by installing turbines under water without dams, looks promising on environmental grounds. So does wave power. A wide variety of devices are under development for both kinds of marine power.[20] At present these technologies are being installed as prototypes (one-offs) and more experience is needed before mass production can be undertaken. The marine environment is tough and not all of the proposed technologies will prove to be durable. Wave and tidal current power can be expected to make significant contributions to electricity supply after 2020.

Natural gas and coal seam methane

Like oil, natural gas is extracted by drilling. *Coal seam methane* is the explosive gas so feared by underground coal-miners that they used to take a canary down with them as a detector. It can be captured from underground mines and utilised. Both types of gas are mostly methane, a greenhouse gas. When methane is burned, it is converted into CO_2, a much less potent global warmer, molecule for molecule.

Electricity generated from natural gas or coal seam methane in a *combined cycle* power station has less than half the CO_2 emissions of a conventional black coal-fired power station and about one-third the CO_2 emissions of a conventional brown coal-fired power station. Gas is even more efficient as a fuel for cogeneration of electricity and heat in commercial buildings and manufacturing industries. It can also substitute temporarily for oil in transportation and for making chemicals, fertilisers, etc. As *peak oil* bites, the demands on gas will increase greatly. However, gas reserves cannot meet all those demands. Already gas is expensive in the Northern Hemisphere and most of Australia's gas reserves are likely to be exported by ship as liquefied natural gas (LNG). If gas alone is used to replace our diminishing supplies of oil, peak oil will be followed within a few decades by peak gas.

Therefore, gas should not be wasted on transportation, but rather be reserved for its most efficient uses: direct heating, cogeneration, trigeneration and as a back-up for renewable energy, especially solar hot water, solar thermal electricity and wind power. In this way, used sparingly, gas can ease the transition to a renewable energy future over the next few decades.

Transport solutions

In using biomass as a source of bioenergy, great care must be taken to avoid adverse environmental impacts, including greenhouse gas emissions. One example is the clearing of tropical rainforests in Indonesia and Malaysia to grow palm plantations to produce palm oil for biodiesel. Both the clearing of forests and the drying of exposed peat lands resulting from that clearing emit greenhouse gases and destroy biodiversity. Another example is the production of ethanol from corn in the USA. This process requires large inputs of fertiliser made from fossil fuels and in some cases even burns coal to distil the ethanol. As a result, the greenhouse benefits are often small or even negligible. The production of biodiesel and ethanol in large quantities inevitably competes with food production.[21]

These mistakes should not be used as a basis for rejecting all uses of bioenergy. As mentioned above, in several countries a large proportion of electricity can be generated from biomass residues, without competing with food production. However, to produce a large proportion of our liquid fuels for transportation from biofuels would be a much more formidable task. The current processes for making ethanol from simple sugars, by means of fermentation and distillation, are inefficient in the sense that they take a large area of land to produce a small quantity of biofuel to substitute for petrol/gasoline. Similarly, the production of biodiesel from oil in crops such as canola or palm trees is very inefficient. These inefficiencies arise because less than 1 per cent of sunlight falling on a hectare of such crops is converted into stored energy. Only a small fraction of that stored energy can at present be turned into useful energy for industrial society. Algae may lift the conversion efficiency to 5 per cent, but more research and development is required. For comparison, some commercial solar photovoltaic modules have efficiencies of 15–20 per cent and this could possibly be increased to 30 per cent by 2015.

Research is under way on so-called second-generation biofuels, in which almost the whole plant is converted into liquid biofuels. High productivity grasses, such as switchgrass, could make a useful contribution if grown on marginal land and processed by second-generation methods. Even so, there is still not enough land to grow enough biomass to substitute for the major proportions of petrol and diesel consumption. At best, biofuels could be used for rural transportation and emergency vehicles. In the face of peak oil production and climate change, this is a significant but limited role. Other alternatives must be sought for transportation. Therefore, a mix of transport modes is recommended.[22]

With more than half the world's population now living in cities, urban public transport, cycling and walking are the key options. Urban public transport can be composed of a mix of different types of transport modes: heavy rail for long distances, light rail and scheduled buses for middle and shorter distances, and demand responsive minibuses with variable routes for taking travellers between home and scheduled, fixed route public transport services. Cycling and walking are important feeders for public transport as well as transport modes in their own right.

Changes in urban density and urban form take place on longer timescales than transport planning. Nevertheless, urban planning must gradually concentrate the population around public transport nodes and stops. Leading transport scholars, Peter Newman and Jeff Kenworthy, have developed a plan for reconstructing an *automobile city* into one that can make much greater use of public transport, cycling and walking. It is based on fostering the existing tendency for large cities to form subcentres, which can be linked by fast heavy rail. The subcentres are the town centres of *transit cities*, each of which has a diameter of 20–30 kilometres. Within each *transit city* are a number of *local centres*, linked to the town centre by light rail or bus. Thus the whole city is encouraged to evolve into clusters of clusters, with public transport planning closely linked to urban form.[23]

For urban residents for whom public transport, cycling and walking are unsuitable, plug-in hybrid and all-electric cars are currently the best solution. Batteries have been improved steadily since the 1980s and promising new designs are coming onto the market. For instance, CSIRO's 'UltraBattery' combines a capacitor and a regular lead-acid battery. The capacitor provides and stores large amounts of electrical power in short periods of time, such as during short acceleration bursts and regenerative braking, while the battery provides a steady electrical output over long periods of time. The technology has been patented by CSIRO, Australia's national research organisation, and is expected to go into mass production in about 2010, to be incorporated initially into hybrid vehicles.[24]

Even when base-load electricity is generated from coal, electric vehicles emit less greenhouse gas than petrol and diesel vehicles. As more and more renewable energy is integrated into the grid, electric vehicles will become less and less greenhouse intensive.

Solutions for agriculture

Within the agricultural sector, the largest sources of emissions are, in order of importance, methane from the digestive processes of cattle and sheep, and nitrous oxide from fertiliser and tilled soil. In addition, agriculture offers large opportunities for *biosequestration* of CO_2 through plant roots into soil. Land-clearing, which is classified separately, is partly caused by agriculture and partly by forestry. Possible technological measures (in the broadest sense) for reducing emissions are:[25]

- Methane emissions could be reduced by selective breeding, herd management, improved nutrition and possibly vaccination.

- Nitrous oxide emissions could be reduced by changing the method, timing and quantity of fertiliser applications; using manure as a fertiliser; and improving soil structure.

- Both sources of emissions could be reduced by reversing the current trend towards increasing meat consumption, especially beef and lamb. This means that industrialised countries will have to set an example for developing countries that are currently moving towards Western diets.

Carbon can be sequestered in soils by the following measures,[26] most of which have multiple benefits:

- Planting trees that capture CO_2, control salinity and erosion, provide shelter-belts for stock, improve water quality and protect biodiversity.

- Improving soil management via minimum tillage, controlling compacting by stock, and maintaining continuous vegetation cover.

- Adding biochar, an organic charcoal made from biomass, which also improves soil fertility and retention of nutrients and moisture.[27]

- Choosing crop species that produce large numbers of phytoliths, microscopic spherical shells of silicon that store carbon in their interiors for thousands of years.[28]

Most of the proposed agricultural measures still require a lot more scientific research.

Sustainable energy scenarios

In practical terms, scenarios are desktop studies that examine the future consequences of different sets of assumptions. They are a means of exploring visions of future society, the potential contributions of various technologies and policies, their economic costs and benefits, and their environmental and social impacts. Provided their assumptions are set out honestly and openly, scenarios are a valuable systematic method of envisioning a problem and its possible solutions. Without visions, we cannot proceed. With unrealistic and hidden assumptions, as in many applications of macroeconomic models to greenhouse mitigation, they can be a means of spreading obfuscation and biased results by vested interests. Here I summarise a small sample of existing transparent scenario studies for greenhouse gas reduction at global and national scales.

Global energy/transport scenarios

Physicist and energy analyst Bent Sørensen has developed global scenarios for energy demand and energy supply in 2050.[29] He considers four energy supply scenarios: 'clean fossil', 'safe nuclear', 'decentralised renewable' and 'centralised renewable'. The first two are deliberately based on hypothetical technological systems that do not exist at present: CCS and a new generation of nuclear power stations, respectively. Although the two renewable energy scenarios involve assumptions about improvements in existing technologies such as off-shore wind farms, PV systems, transmission lines and energy storage, they are in my view much closer to reality than the first two and so are of particular interest for this book.

The 'decentralised renewable' scenario explores how far one can go with residential systems (small-scale solar, wind and fuel cells located in individual households). It is limited by the fact that renewable energy resources are distributed inequitably over the world.

The 'centralised renewable' scenario is still decentralised in comparison with the 'clean fossil' and 'safe nuclear' scenarios. While keeping many decentralised energy systems such as residential solar, it places some types of renewable energy system on non-arable land and off-shore and transmits the energy to consumers by transmission lines or pipelines. Its energy mix comprises energy efficiency, bioenergy, wind power (both on-shore and off-shore) and solar power. No additional hydroelectric power is included beyond plant that is existing and under construction. All the scenarios use a geographic information system to assess the extent to which renewable energy resources match energy demand in different regions of the world. In regions where there is a poor match between supply and demand, import and export of energy via transmission line and pipeline are added. The results are particularly encouraging for the centralised renewable scenario: there is in total a global oversupply of renewable energy potential and good matches between supply and demand can be achieved in all regions. Furthermore, food production is not compromised by using biomass residues to produce bioenergy.[30]

The decentralised renewable scenario has much greater difficulties in matching supply to demand in different regions. It would require a much larger intercontinental trade in energy than the centralised renewable scenario, an outcome that appears to contradict the original motivation for decentralised energy. On a global scale, the decentralised renewable scenario leaves little room for increases in supply.

In 2009, McKinsey & Company, supported by ten leading companies and organisations, published a revised global cost curve for global greenhouse gas abatement in 2030.[31] A cost curve ranks different abatement options according to cost, while showing the quantity of abatement that could be achieved by each option at its own cost level. For example, it might show that switching from incandescent to LED lighting in the residential sector could save in 2030 about 200 Mt per year of CO_2-equivalent emissions at a cost saving of €95 per tonne (€/t), while 'low penetration' wind power might save in 2030 about 2 Gt per year of CO_2-equivalent emissions at a cost of €12/t.[32] Cost curves can be produced for the present time or, in the case of the McKinsey report, for a future date. The assumptions made in projecting future costs are spelled out clearly. The study is limited to technical abatement measures costing less than €60/t. It assumes no change in lifestyles and continuing increases in global prosperity over the 21-year timescale. It does not address the policies needed to achieve the technical potential. In broad terms, the principal results confirm and extend earlier work by a wide range of authors, including McKinsey:

- There is potential to reduce global greenhouse gas emissions to 35 per cent below the 1990 level by 2030.
- This reduction would 'have a good chance' of keeping global warming below a 2 degree Celsius increase.
- A delay of 10 years would make it 'virtually impossible' to keep below a 2 degree Celsius increase.
- Future economic savings from efficient energy use could pay for most of the additional up-front investment in cleaner energy supply and other infrastructure.

Although recent science suggests that a 2 degree Celsius increase in global average temperature is unsafe, this study is still very useful in identifying the 'low-hanging fruit' and the potential emission reductions to be gained from various measures at various costs. Debate is inevitable over details – for example, I disagree with McKinsey's assumption that new nuclear power will be a little cheaper than wind power in 2030, when it cannot compete with wind power in 2009 in the UK and USA, and I'm sceptical that new coal power with CCS will cost as little as €30/t in 2030. Nevertheless, the principal qualitative results listed above appear to be robust.

More detailed scenario studies have been carried out for several countries and groups of countries. We look at a few of these now.

Individual country scenarios

In the USA, two recent studies are of interest. In 2007, engineer and energy analyst Arjun Makhijani published an energy technology scenario for USA in 2050 that is free of both fossil fuels and nuclear power. The author points out that his scenarios avoid the triple 'energy insecurities' of global warming, peak oil and price insecurity of fossil fuels, and nuclear terrorism and proliferation. Without doing a detailed economic analysis, Makhijani points to the enormous cost savings potentially available from energy efficiency and argues, as others have, that they can *offset* a large proportion of the additional costs of renewable energy.[33]

Another engineering study for the USA, by Mark Jacobson, considers global warming, air pollution and energy security in ranking energy supply and transport technologies according to a wide range of risk-weighted impacts. He finds that wind, solar thermal electricity, geothermal, PV and marine power could provide all of US electricity for both stationary energy and electric vehicles, and that coal with CCS and nuclear power 'offer less benefit and thus offer an opportunity cost loss and the biofuel options provide no certain benefit and the greatest negative impacts'. Furthermore, 'the sound implementation of the recommended options requires identifying good locations of energy resources, updating the transmission system, and mass-producing the clean energy and vehicle technologies'. He does not attempt to estimate a timescale for this radical but essential transition.[34]

In Australia, despite its very high per capita greenhouse gas emissions, several scenario studies suggest that emissions could be cut by

30–40 per cent by 2020, given appropriate policies across the board in energy, transport, non-energy industry, agriculture and forestry. The different targets achieved depend on assumptions about how much solar and geothermal would be commercially available by 2020, whether significant changes in diet could be achieved and whether aluminium smelting is required to become greenhouse neutral.[35]

China has huge potential for both unsustainable and sustainable development. On one hand, China is the world's largest producer and consumer of coal and recently overtook the USA as the world's largest emitter of greenhouse gases. Emissions from electricity generation (mostly coal) and construction are high and growing rapidly, together with sales of motor vehicles. Nevertheless, China's average per capita emissions are still relatively low.

On the other hand, China is well endowed with sunshine, wind and biomass residues, and has set a target of 15 per cent of all energy consumption (not just electricity) from renewable sources by 2020. China is the world leader in the production and use of solar hot water systems. It also overtook Japan in 2008 to become the biggest producer of solar *photovoltaic cells*, although the vast majority are exported. From 2006 to 2008 inclusive, China doubled its wind power capacity each year, so that by the end of 2008 it had the fourth largest in the world. China is developing geothermal energy for heating, but has not yet developed high-temperature sources for electricity. Urban railways are growing rapidly in major cities such as Shanghai, Beijing, Shenzhen and Chengdu. Motor vehicles manufactured in China have to meet fuel efficiency standards 40 per cent higher than the USA's.[36]

At present Chinese energy policy envisages energy efficiency and renewable energy supplying a large fraction of its *growth* in energy consumption. As someone who travels frequently to China, I believe that the next step, of increasing the use of these sustainable energy technologies in China to the extent that they substitute for part of *existing* fossil fuel use, depends to some extent on the example set by Western countries. China's unsustainable development pathway is modelled on the Western (especially USA's) development pathway.

Those in China promoting a sustainable development pathway, which could leapfrog over the mistakes of the West, would gain strength and influence from stronger reductions in Western emissions. To further encourage China's sustainable development pathway, detailed scenario studies are needed for its renewable energy futures.

An integrated renewable electricity supply scenario

Electricity generation, mainly from coal, and transport, mainly from oil combustion, are the two largest single contributors to global greenhouse gas emissions. Since oil consumption will inevitably decrease as the result of peak oil, and one of the best current substitutes is electric transportation, it is worth focusing on the future of electricity generation.

Renewable sources of electricity offer a diverse group of technologies at various stages of development:

- medium-cost commercial (on-shore wind; landfill gas; bioelectricity from the residues of plantation forestry and agricultural crops);
- expensive commercial (conventional off-shore wind; proven types of solar thermal electric; conventional solar PV; bioelectricity from dedicated crops and dedicated plantation forests);
- prototypes (new solar PV; some types of solar thermal electric; wave and ocean current; hot rock geothermal);
- experimental (off-shore wind on floating platforms).

This mix of renewable energy technologies will evolve over time, depending upon government policies to foster technological improvement and innovation, market growth and infrastructure such as transmission lines. Given appropriate policies, between 2009 and 2020, the main reductions in greenhouse gas emissions from the electricity sector will come from energy efficiency, solar hot water, wind power and bioelectricity from residues. Contrary to the propaganda from vested interests, energy efficiency and all of these renewable sources of electricity can contribute to base-load power, as discussed under fallacy 8 in chapter 2.

During 2009–20, further development of the other categories of renewable electricity technology will take place. Shortly before 2020, we are likely to see large and rapidly growing contributions from solar thermal electricity, hot rock geothermal power and solar PV, all of which will continue to expand through to 2030. The first two of these sources are also base-load. During the 2020s, wave and tidal current power may become significant contributors. By the 2020s, it's possible that low-cost electrical storage will be available.

There is no technological limitation on most countries generating all their electricity from renewable sources by 2050 or thereabouts. The mix of sources will vary with geographic region and time. Some trade in energy will be required, as it is today, to balance supply and demand. The tropics, for example, are not generally suitable regions for much wind power or for solar concentrators, although they have potential for flat-plate solar collectors for both hot water and residential PV electricity generation, bioenergy and in some places marine and geothermal power. Europe has much potential for wind in the north, hot rock geothermal in several locations, and solar in the south – the latter could be greatly expanded with solar electricity transmitted from North Africa. The USA has separate regions high in wind, solar, biomass and geothermal potential. Australia has wind in the south, solar in the north and hot rock geothermal in the centre.

Treating climate action like a wartime emergency

Climate change activists such as Lester Brown[37] and David Spratt and Philip Sutton[38] claim that the military mobilisation by the United States during World War II serves as a useful example of production and technologies being changed rapidly on a huge scale. While I agree that we have something to learn from wartime experience, it's not as simple as it sounds. Drawing an analogy between mobilising for a war and mobilising against human-induced climate change contains a weakness, as Patrick Hodder and Brian Martin from the University of

Wollongong explain in a personal communication:

> In a war, the very survival of governments is directly and immedi-
> ately threatened. Governments therefore have a vested interest in
> leading an emergency response to the threat of war. By contrast,
> climate change does not immediately threaten governments in the
> developed world. These governments do not have an interest in
> leading an emergency response to climate change. Indeed, some
> such as the Australian federal government and state governments
> in Victoria, New South Wales and Queensland have a vested inter-
> est in maintaining business-as-usual and even helping promote
> and expand the coal industry. Governments with an interest in the
> current business paradigm have little incentive to fundamentally
> change the economy. Governments will not engage in an emergency
> response to climate change until they are convinced, possibly by a
> social movement, that the failure to lead on climate change repre-
> sents an immediate and direct threat to their survival.

Another difference is that conducting a war generally involves relatively simple and small changes to the economy.[39] The key tasks are
to make large quantities of weapons while maintaining most of the
economy in its pre-existing state, apart from a pause in the production
of luxury goods and services. The motor vehicle industry can be readily transformed into manufacturing tanks and other military vehicles;
the aircraft industry expands while replacing civilian with military
planes; and hand-weapons are readily produced by modifying existing
factories. In the economic-industrial sphere, many winners and few
losers emerge from war.

On the other hand, emergency climate action is very different
from the slow transition scenarios studied in the Stern and Garnaut
reports. It involves radical transformations of the energy, manufacturing, agricultural and forestry sectors. Coal-fired power stations cannot
be converted into wind and solar farms or even easily into combined-
cycle gas-fired power stations. Energy consumption must decline
rather than expand. Existing motor car production lines cannot be

readily modified to manufacture trains or even electric vehicles. The demand for aluminium will fall as renewable electricity at three or four times the price replaces dirt-cheap subsidised coal power for smelting. Cement making will require entirely new chemical processes that are still at the developmental stage. Native forest logging must be replaced with plantation forestry using different technologies. These industries will experience stranded assets on a huge scale. It is not surprising that they and their representatives in government are resisting strong climate action.

Even without such resistance, the emergency transformation will take much longer than preparing a country for war. Professionals, such as electric power engineers, and technicians must be trained in great numbers; factories must be built; trade unions and other interest groups must be convinced; and technologies must be properly tested (unlike many weapons in wartime) before being mass-produced.

Recognising these political and economic/industry barriers does not constitute a logical argument against striving for emergency climate action. Climate science shows that we are facing an emergency and therefore the climate action movement must work on this basis to win over the community at large and governments. However, its members must do this with eyes wide open, not deluding themselves that the task is as simple as making war.

Conclusion

We already have the technologies and other measures for replacing coal and oil with zero-emission sources: ecologically sustainable energy comprising energy efficiency and renewable energy. These measures, taken together, are safe and reliable. Those that reduce the demand for energy – energy efficiency, energy conservation and, in many locations, solar hot water – will actually save money for consumers. If we become a smart society, we can ensure that the economic savings from demand reduction pay for a large fraction of the additional costs of renewable energy.

All the renewable energy supply technologies are currently more expensive than burning coal or oil without CCS. But that is not a fair comparison, since it excludes the environmental, health and economic damage done by burning these fossil fuels. Some renewable energy sources – wind power and the low-cost forms of bioenergy from residues – are less expensive than nuclear power and some projections of the cost of coal power with CCS. That is a fair comparison, provided that we take into account the additional benefits of renewable energy: notably reduced air and water pollution, reduced water consumption and energy security. Anyway, as pointed out in fallacy 5 of chapter 2, conventional nuclear power will become a significant greenhouse polluter within several decades, so it should not be considered as an option.

In the transition to a sustainable energy future, the least polluting of the fossil fuels, natural gas and coal seam methane, can play a valuable role in assisting the phase-out of dirty coal power, especially by cogeneration and backing up renewable energy systems. However, it would be a mistake to give all the incentives to such a limited resource as gas. Policies to expand energy efficiency and renewable energy must be the top priority now, even if the latter technologies are a little more expensive than gas in the short term. To address the climate crisis, we must plan on a 40-year timescale at least. We must start strongly now and achieve big reductions in emissions before 2020. But we cannot expect to completely transform our energy technologies in a decade.

Finally, we must recognise that technologies alone will not completely solve the climate crisis. We must insist that our governments also put in place processes and policies to transform economies into steady-state systems and to stabilise population numbers. These non-technological approaches are discussed, together with many other policies, in the next chapter.

4

ESSENTIAL POLICIES
FOR THE
21ST CENTURY

Emissions trading schemes are sometimes touted as the new panacea for climate change. The European Union has had an emissions trading scheme since the beginning of 2005, the Australian and US governments are proposing to introduce schemes in the near future. However, there is growing concern that these schemes may be ineffective. In practice governments can shape the market and lifestyles by means of a wide range of policy instruments apart from emissions trading: a carbon tax; laws, regulations and standards; institutional change; fees, grants and subsidies; education and information; and support services. The principal goal of the climate action movement must be to gain a broad portfolio of effective policies at all levels of government. It can do this by bringing pressure to bear directly on governments and by mobilising the community at large. While some form of carbon price, via an emissions trading scheme or a carbon tax, is necessary, it is far from being sufficient.

A policy is a statement of intention. To be effective, it must be coupled with a strategy and process for implementation and enforcement. This chapter reviews the essential policies needed from governments on national, state and municipal (whole city) scales to achieve large reductions in greenhouse gas emissions.

Although in theory big business and industry operate within a market shaped by government, in practice they strongly influence the government policies that shape the market. In some cases, they even write those policies (see chapter 2). They do this by using their huge, concentrated wealth to lobby, make political donations, buy into the media, fund public relations campaigns and advertising, set up 'think-tanks', initiate lawsuits, and produce and disseminate 'educational' materials.[1] Therefore, the climate action movement must also seek to change the policies, or at least the norms, of business. However, my principal policy focus here is on governments.

Key government policies needed

Greenhouse gas mitigation is a complex problem involving a transformation of technologies, economies, societies, lifestyles and governance. Contrary to the dogma of economic fundamentalism, a single 'silver bullet' policy, such as a cap-and-trade emissions trading scheme, can't achieve these transformations on its own – see fallacies 13 and 14 in chapter 2. There is a growing consensus within the climate action movement about the wide range of basic policies that would achieve an effective greenhouse mitigation strategy. In broad terms, we need:

- Science-based national greenhouse gas emissions targets, both short-term and long-term, to set the direction of change.

- An international target for atmospheric CO_2 concentration of 350 ppm or lower.

- Targets, both short-term and long-term, for renewable electricity, renewable heat and efficient energy use. The short-term targets are necessary for building up the industries for the only carbon-free energy technologies that are commercially available or very close to that stage.

- A ban on all new conventional (that is, greenhouse polluting) coal-fired power stations, in order to meet all scientifically-based greenhouse targets and to foster efficient energy use. Major refurbishments and expansions of existing coal-fired power stations, and oil produced from coal and tar sands, should be included in that ban.

- A significant carbon price that increases over time.

- Termination of subsidies to the producers and users of fossil fuels.

- Several additional policies to expand markets for renewable energy until the carbon price is high enough to take over. These policies could include gross feed-in tariffs, tax concessions and renewable energy portfolio standards (discussed below).

- Regulations and standards for efficient energy use in all buildings and energy-using appliances and equipment.

- A package of measures to foster a socially just transition for low-income earners and workers in greenhouse-intensive industries.

- An international agreement to set the nations on the pathway known as 'Contraction and Convergence', with the goal of achieving the same average per capita greenhouse gas emissions by all countries within several decades.

- Essential infrastructure – such as railways and transmission lines – to enable the transition to an energy efficient, renewable energy future.

- Research and innovation, especially for efficient energy use, renewable energy, biosequestration of CO_2, cleaner industrial processes, new transport technologies and a steady-state economic system.

- Termination of the logging of native forests and land clearing for agriculture.

- In agriculture, policies to reduce emissions, especially by cutting the quantity of beef and lamb consumed, and to encourage biosequestration.

- Increased education and training to provide the skilled workforce for the transition.

- Levelling of population growth, especially in countries that have very high per capita greenhouse gas emissions, such as the USA and Australia.

- Development and implementation of a steady-state economic system, that is, one that incorporates limits to the growth in the use of energy, materials and land.

At first glance, this list looks formidable. However, some of these policies could be implemented quickly at low cost: a ban on new conventional coal-fired power stations and on the logging of native forests and land clearing, together with the introduction of measures for a socially just transition, would achieve almost immediate results. Other policies, such as the last dot point, could take half a century to be fully implemented. The remaining points could take one to two decades. However, all of the proposed policies must be commenced now. Thus the policy list gives a program to benefit society and the environment over the next two generations. I now examine each policy in more detail.

Set strong greenhouse targets

At this stage of the climate crisis, we need national targets for greenhouse gas emissions and an agreed international target for CO_2 concentration in the atmosphere. Both kinds of targets should be based upon the latest scientific knowledge.

According to the 2007 Intergovernmental Panel on Climate Change (IPCC) scenarios, in order to stabilise CO_2 concentrations in the atmosphere at 350–400 ppm (a CO_2-equivalent concentration of 445–490 ppm), global CO_2 emissions would have to peak by 2015 and decline to 50–85 per cent below the 2000 level by 2050.[2] Since the wealthy industrialised countries are collectively responsible for the vast majority of greenhouse gases accumulated in the atmosphere after the Industrial Revolution, they should be morally required to meet stronger national targets than the global average, namely reductions in the range of 80–100 per cent by 2050. Even the CO_2 target range of 350–400 ppm is regarded as risky by leading climate scientist James Hansen, who recommends an initial atmospheric CO_2 concentration target of 350 ppm, keeping open the possibility of further reductions below this level if required.[3] The actual 2008 CO_2 concentration was 385 ppm.

Because most governments have much shorter periods in office than 40 years and because greenhouse mitigation becomes more expensive the longer it is delayed, short-term targets are essential as well. At the December 2007 United Nations climate conference in Bali, the governments of several nations proposed a short-term target of a reduction in total CO_2-equivalent emissions of industrialised countries of 25–40 per cent below the 1990 level by 2020. As indicated by the scenarios discussed in chapter 3, a reduction of 30 per cent below the 1990 level by 2020 is feasible, based on existing technologies, for Australia,[4] the biggest per capita emitter of the industrialised world. Therefore, it should be feasible for most other developed countries. Modest technological improvements in solar and geothermal power should be able to increase this reduction to 40 per cent.[5]

Table 4.1 shows the unconditional greenhouse targets in February 2009 of several countries and groupings of countries. It is important to have both a long-term (2050) vision and a short-term (2020) commitment, with the policies set in place to achieve both. In my view, developed countries should be setting 2020 targets of at least 25 per cent below their 1990 levels of emissions and initial 2050 targets of at least 80 per cent below their 1990 levels. Climate movement organisations should be pushing their governments for stronger targets than these minima.

TABLE 4.1

Greenhouse targets extant in April 2009

Country or grouping	2020 reduction target	2050 reduction target	Comparison year for reduction
EU	20%	–	1990
UK	at least 34%	80%	1990
USA (Obama pre-election promise)	0%	80%	1990
California	0%	80%	1990
Mexico	–	50%	2002
Australia	5%	60%	2000
Annex 1 countries (proposed at Bali December 2007, but not agreed to)	25–40%	–	1990

SOURCES Australian Government; Obama & Biden; EU news; AAP; websites of Departments of Climate Change from several countries and states.[6]

Set targets for energy efficiency, renewable electricity and renewable heat

Efficient energy use and renewable energy are the only carbon-free[7] energy options that are commercially available or close to that stage. They are generally much safer and less complex than fossil fuel and nuclear technologies. Another advantage is that they can create more local employment per unit of energy generated or saved than fossil and nuclear technologies (see chapter 2, fallacy 12). Despite these clear advantages until recently, they have been undervalued and neglected in most countries. For all these reasons, they deserve special support at this critical period of global warming. Targets are an essential part of that support.

Table 4.2 shows existing targets for 2020. These targets are feasible, but difficult. Renewable heat is shown separately, because of growing awareness that heat is a significant source of emissions from industrialised countries and that renewable heat has not received sufficient attention in comparison with renewable electricity. My own scenario studies found that Australia could increase its renewable energy contribution to electricity generation from about 8 per cent in 2004–06 to nearly 25 per cent by 2020, given the political will.[8]

TABLE 4.2

Renewable energy and energy efficiency targets for 2020 extant in April 2009

Country or grouping	2020 renewable energy target	2020 energy efficiency target	2020 renewable heat target
EU	20% of all energy, not just electricity[9]	20% improvement	included in 20% of all energy
UK	20% of electricity[10]	–	–
USA[11]	25% of electricity by 2025	15% reduction of electricity demand below DOE's projected level	–
Australia	20% of electricity[12]	–	–

SOURCES See 'Key readings and websites': Greenhouse mitigation: scenarios, economics and policies.

Al Gore's inspiring program – Repower America – aims to transform US electricity to 100 per cent 'renewable energy and truly clean carbon-free sources' within a decade.[13] While 100 per cent renewable energy is indeed feasible for the USA, Europe, Australia and many other countries, ten years is too short a time for such a dramatic transformation in countries that are still highly dependent upon fossil fuels. Currently the USA, the EU and Australia each generate less than 10 per cent of their electricity from renewable energy and the UK less than 5 per cent. Much of the renewable energy comes from hydroelectricity, which has little additional potential in these countries. Within the next decade, the principal renewable electricity contributions will have to come from the technologies for which there is greatest commercial experience: wind power and bioelectricity. Solar electricity (both PV and solar thermal), and hot rock geothermal power, both show huge potential, but neither of these technologies is ready for rapid scale-up, as discussed in chapter 3. Furthermore, both thermal and electrical storage need further development and testing. Several years will be needed to discover which of the many new competing solar thermal systems will prove to be commercially viable. Some of these systems are still in the prototype phase of development. Hot rock geothermal is just beginning to generate tiny quantities of electricity. In addition, it will be necessary to build the factories to manufacture the new technologies, to rebuild transmission systems to carry a much more geographically distributed set of power stations, and to educate and train the engineers and tradespeople to build, maintain and install the new systems. Over 2006–08, even in the absence of a challenging renewable energy target, supply could not meet demand for wind turbines and high-grade silicon for solar PV cells.

We need to give greenhouse mitigation the priority of a wartime emergency (see chapters 1 and 3), but even so a complete transformation of the electricity-generating system will take longer than 20 years. It does not help to underestimate the magnitude of the task ahead.

Ban new conventional coal-fired power stations

Until carbon prices are sufficiently high to make the cost of new conventional coal-fired power stations prohibitive, the climate action movement should push governments to enact explicit policy to ban new dirty coal power. This is urgently required for two reasons:

- Every conventional coal-fired power station emits typically 6–12 million tones of CO_2 per year for up to 40 years. Each monster that is brought into operation makes the task of cutting emissions more difficult and more expensive.

- Every new base-load power station undermines energy efficiency programs. This is because power station developers must ensure that they can repay the huge capital costs from electricity sales. Their nightmare is to commission a new power station as electricity demand is decreasing. So they generally work closely with government and industry on programs to encourage demand growth.

Phasing out coal is the key to drawing Earth's climate away from tipping points, to a safe state. James Hansen shows that a prompt ban on all new coal-fired power stations on this planet, coupled with a phase-out of existing coal plants linearly over the period 2010–30, would lead to a peak in atmospheric CO_2 concentrations during the next few decades at 400–425 ppm. Thereafter, CO_2 concentrations would gradually decline, with the rate of decline depending upon the rate the remaining oil and gas reserves were used.[14]

The developers and supporters of some new coal-fired power stations are claiming that their stations will be 'CCS-ready', that is, their design would allow carbon capture and sequestration to be easily added at a future date.[15] The climate action movement should not accept that claim as a reason for avoiding the ban. CCS is not commercially available and it is very unlikely that it will become so in the near future.

Needless to say, in the absence of CCS, we should insist that the ban should also be applied to major renovations and extensions to existing coal-fired power stations and to other highly greenhouse-intensive sources of fossil fuels, such as oil produced from coal, shale and tar sands.

California has set an excellent precedent: a temporary Greenhouse Gas Performance Standard that in effect bans new power stations with CO_2 emissions intensity greater than that of a combined cycle natural gas power station.[16]

Impose a carbon price that increases with time

A *carbon price* is a vital policy instrument for encouraging shifts in the economic structure towards systems that are less greenhouse intensive. A carbon tax and an emissions trading scheme (ETS) of the cap-and-trade type are the two principal methods for doing this. A carbon tax specifies the price of carbon emissions of various types and then allows the market to determine the quantity of emissions reduced. In contrast, an ETS specifies the reductions in emissions required at various times in the future and then supply and demand in the artificial market for emission permits determines the price of these reductions. This section does not attempt to give a detailed description of each instrument, but rather highlights the key issues that determine the effectiveness of each.

Carbon tax

A carbon tax is a tax on the carbon content of fossil fuels. It is the simpler of the two schemes to implement and operate, and is more difficult to cheat on. Most countries already have government departments that manage taxation. Part of the taxation revenue received could be returned to households as a monthly dividend, with equal shares on a per capita basis (half-shares for children up to a maximum of two child-shares per family). A carbon tax with dividend is progressive:

a family with a large house and several large cars will have a tax that is much greater than the dividend; on the other hand, a family reducing its greenhouse gas emissions to less than average will make money. The remainder of the tax, that is not transferred to the dividend, can be used to assist low-income earners and workers who are displaced by the transition (see below).

The principal potential shortcomings of a carbon tax are twofold. Firstly, powerful industries, that may be big greenhouse gas emitters, may be able to obtain exemptions from the tax. This is the case in several European countries, where export industries are either exempt or have lower tax rates than domestic industries.

Secondly, if people have no alternative technologies or measures, then they simply have to pay the tax and their emissions won't change. Clearly a carbon tax must be accompanied by government policies to support the construction of the necessary infrastructure, to make legal and institutional changes, to support complementary measures and to fund research and development and education and training of the workforce. Both these shortcomings can be readily overcome by appropriate government policies.

Despite, or perhaps because of, the advantages of a carbon tax, corporate interests have been successful in lobbying many governments against them. They have created the false impression that a carbon tax would necessarily cost consumers more than an ETS. Lobbyists for big business have found it easy to incorporate their campaign against a carbon tax into their pre-existing campaigns against all new taxes.

Emission trading scheme

An ETS is a complicated means of establishing a carbon price. A cap-and-trade ETS is set up in the following way. Caps, that is, limits on total future emissions from certain sectors of the economy, are announced by the government. Emission permits are issued to the largest greenhouse gas emitters, with the total number of permits corresponding to the cap on emissions from those sectors. By the end of a particular period, the emitting companies must surrender permits equivalent to the quantity of emissions from their installations over that period. Provided the cap is lower than the current level of emissions, permits are scarce and so acquire a market value. The permits are tradable.

Consider a food processing business that emits so much greenhouse gas that it must have permits. To the extent that it can reduce its emissions at a cost less than the market price of permits (for example, by energy efficiency and solar process heat), it will do so and will sell its excess permits to another big emitter, say an aluminium smelter, that cannot. Large industrial facilities that can only reduce their emissions at costs that are greater than the market price of permits will have to purchase permits to cover their emissions. Thus the lowest cost measures for reducing emissions are implemented first, a process that is in theory *economically efficient*. Also, the market sends a message to investors that any new aluminium smelter will have to be designed to be much more energy efficient than the old one. The carbon price flows through the whole economy, making all greenhouse-intensive products covered by the scheme more expensive than before. In the present example, aluminium cans would become more expensive, encouraging substitution of (say) recyclable glass bottles. As the carbon price rises over years and decades, the economic structure of the country or region where the ETS is applied gradually changes to one that is less greenhouse intensive.

Businesses and politicians consider an ETS to be a market measure while a carbon tax is labelled as a regulatory or so-called 'command and control' measure. However, in practice, both a carbon tax and an ETS require a legal framework with enforcement and so both are mixes of market and regulatory measures.

The key requirements for an effective and fair ETS are listed below.

1 Allocation of permits by auction, not grandfathering
All the emission permits should be auctioned. However, industry has lobbied for *grandfathering*, in which permits are allocated free of charge to the big emitters according to their current or previous levels

of emissions. The disadvantages of grandfathering are:

- Once the permits have been allocated within the cap, they acquire a market value, whether they were grandfathered or auctioned, and this price is passed on to consumers, pouring a huge windfall profit into the pockets of industries that receive them free of charge, thus violating the polluter pays principle.

- New, cleaner industries will have to buy permits from existing industries that received them free of charge, disadvantaging the newer industries in the competition and so slowing the necessary transition in economic structure.

- Grandfathering provides a perverse incentive for big emitters to increase their emissions before the allocation dates.

- The revenue raised by auctioning permits can be used to assist low-income earners and workers who are disadvantaged by the carbon price and also to fund the new infrastructure needed for the transition to a low-carbon future. With free allocation, there is no revenue.

The European Union (EU) ETS covers CO_2 emissions from stationary energy and industry. In the first phase of the EU scheme, 99.8 per cent all emission permits were allocated free of charge and in phase II (2008–12) about 97 per cent. However, in phase III (2013–20), 100 per cent of emission permits for electricity generators and at least 80 per cent of all permits will be auctioned. Free permits will still be issued to emissions-intensive European-based industries, such as cement and steel, which have to compete in world markets.

The US Regional Greenhouse Gas Initiative (RGGI),[17] an ETS, commenced on 1 January 2009, after early auctions in 2008. The ten US states in the RGGI have agreed to 100 per cent auctioning and so has President Obama for his proposed national ETS.[18]

The Australian ETS, called the Carbon Pollution Reduction Scheme, appears superficially to be better than the EU ETS in this regard, in that initially 70–76 per cent of permits will be auctioned. However, this improvement is negated to a large degree by the government's promise to give most of the free permits to the biggest green-house polluters. Part of this is presented as so-called 'compensation' to coal-fired electricity generators under the *Electricity Sector Adjustment Scheme*.[19] Assuming a CO_2 price of A$25 per tonne, the strategic value advisors Innovest estimate that these generators will receive permits of about A$654 million in 2010–11 yielding a total allocation of A$3.9 billion over the first five years of the scheme. The lion's share of this windfall will go to the brown coal electricity generators in Victoria, who own the most greenhouse-polluting power stations in Australia. In addition, so-called emissions-intensive trade-exposed (EITE) industries will receive an estimated A$2.8 billion in free permits in 2010–11 alone, rising to A$4.4 billion in 2014–15. The lion's share will go to aluminium smelting. The majority of the payout will go to overseas-owned companies.[20] To make matters worse, provision has been made for the free allocation to EITE industries to expand with time up to 45 per cent of all permits by 2020, a clear incentive to increase emissions.

The case against 'compensation' is made in the Garnaut Report (see 'Key readings and websites') and also in box 4.1 by electrical engineer Iain MacGill and economist Regina Betz from the University of New South Wales. A fairer method is to require EITE industries to purchase auctioned emissions and apply *border tax adjustments* to exports and imports. This will also maintain the effectiveness of the emissions trading scheme.[21]

BOX 4.1

The case against compensation of big emitters in carbon pricing

Iain MacGill and Regina Betz of the Centre for Energy and Environmental Markets at the University of New South Wales[22]

In practice, compensation will also have an impact on the cost-effectiveness of the emission-reducing scheme. Firstly, it puts the focus of industry players on trying to maximise their compensation rather than finding new and cheaper ways to reduce their emissions, encouraging the 'victim mentality'. Even more importantly, compensation for large emitters risks undermining the good governance that is essential to deliver an effective ETS.

'Compensation' is generally understood to mean righting a wrong or imposition placed upon some party. Putting a price on emissions doesn't represent an additional imposition on emitters, but rather, the removal of a public subsidy. Emitters have been knowingly receiving this subsidy since at least the Rio Declaration [on Environment and Development] in 1992. The public actually has a pretty good case for reparations from those industry players that played a role in delaying action to end this subsidy through their lobbying and other efforts.

Furthermore, arguing that introducing an ETS without compensation will have an adverse impact on investor confidence in good governance has it the wrong way around. Large emitters are typically owned by investors through their shareholdings. Most investors over the last decade have made the judgment that climate change is a problem, and that 'polluter pays' policies were coming. Presumably, some other investors have judged that there wasn't a problem or, worse, that governments would inevitably yield to corporate demands for so-called compensation. Paying such compensation therefore rewards the wrong group of investors – those taking a bet against good governance.

As for industry leaders, governments should be supporting those trying to do the right thing. If good corporate citizens see others being rewarded for claiming victim status, they are almost obliged to attempt the same thing. After all, when governance is weak you'd better be at the table in Canberra or you'll probably end up on the menu.

2 High and increasing carbon price

Another key requirement for an effective ETS is to start with a carbon price that is high enough to initiate changes in economic structure (in particular in the energy supply industry to make it less greenhouse intensive). Then the cap on emissions should be tightened (that is, the allowed total amount of emissions reduced) in steps every few years.

This would push up carbon prices and so bring into the economy the more expensive, cleaner technologies.

In the trial phase of the EU scheme (2005–07), the carbon price peaked at around €26 per tonne of CO_2 in the first year of its operation. In late April 2006, the first public release of emissions data revealed that some governments had allocated more permits than they had emissions. This caused the price of these permits to plummet from €30 to €20 to €10, then to zero over the next year. In phase II of the scheme (2008–12), allocation was tightened up and in 2007 phase II permits traded initially between €12 and €25 per tonne[23] and peaked at €30 in 2008. However, in February 2009 the carbon price of phase II permits dropped below €10 per tonne, apparently as a result of the financial crisis and the ensuing drop in demand for energy. An effect of this fluctuating and generally low carbon price is that, despite the ETS, European countries are planning on developing many new conventional coal-fired power stations.[24]

In Australia, with a very low unconditional target for reducing emissions, the initial carbon price in 2010 is expected to be about A\$23 per tonne of CO_2. This may be too low to drive any significant changes in electricity supply, apart from possibly replacing some brown coal power with black. (See also discussion of fallacies 13 and 14 in chapter 2.)

3 Broad scope

An effective ETS must have sufficient scope in the economy to actually reduce total emissions from the country or region. The EU scheme, for example, is limited to stationary energy (mostly electricity and heat) and some non-energy industrial emissions and to the principal greenhouse gas, CO_2. It covers about half of the EU's CO_2 emissions and about 40 per cent of all EU greenhouse gas emissions covered by the Kyoto Protocol.

The Australian scheme is broader, covering all six greenhouse gases that come under the Kyoto Protocol plus energy generation, transport, *fugitive emissions*, industrial processes, waste and part of the forestry[25] sector. For at least the first three years of the scheme, motorists will be shielded by means of reductions in the fuel tax. Agriculture may be included at a later date, although it does not readily fit into an ETS and so may require different instruments to reduce emissions and increase biosequestration.[26]

4 Emission permits should be temporary licences

It is important, some would say essential, that emission permits are not permanent property rights, but are rather temporary licences to emit. Otherwise, if climate change accelerates and governments have to buy back permits from industry, it could become very expensive for taxpayers.

In the EU scheme, emission permits are temporary licences, but in the Australian scheme they are permanent property rights. It seems that the Australian Government learned nothing from the current drought on the Murray-Darling river system, where it had to buy back water rights from the irrigators of cotton and rice farms.

5 Limit offsets from overseas emission reductions

Both the Kyoto Protocol and the European and Australian ETSs make provision for overseas offsets, that is, gaining credit for emission reductions by reducing emissions in another country or region. Under Kyoto, three categories of international offsets, known as *flexibility mechanisms*, exist.[27] Firstly, there is *International Emissions Trading*, that is, trading of greenhouse gas emission permits on an international market. At present this is only done within the EU. Secondly, if an industrialised country reduces its emissions via a project in another industrialised country, the mechanism is known as *Joint Implementation*. Thirdly, if an industrialised country reduces its emissions by means of a project in a non-industrialised country, it comes under the *Clean Development Mechanism (CDM)*. In the latter case, funds are transferred from the developed to the less developed country, which makes it attractive to the latter. However, there are growing concerns that many projects conducted under CDM would have been done anyway as part of the development process and are not really additional. Further critiques are based on the environmental and social impacts of large hydroelectric projects receiving CDM support.[28]

A developed country or industry in a developed country can use these mechanisms to avoid the necessary economic restructuring at home. This damages the credibility of the greenhouse mitigation strategies of developed countries. Thus, the likelihood of bringing less developed countries such as China and India into an international agreement with mandatory greenhouse targets is reduced. Therefore offsets should be limited to being a small fraction of a country's greenhouse target. Yet phase II of the EU scheme allows a large fraction of its emissions reductions under the ETS to come from offsets. The Australian scheme has no limit on international offsets. This in effect limits the pressure from the carbon price to encourage energy efficiency[29] and also bodes ill for effective international agreements.

6 Strong penalty and make-good provisions
If in a given time an emitter exceeds its emissions quota, as determined by the number of permits it possesses, it should be required to pay a large penalty and have its emissions quota in the next period reduced by the excess. This should ensure that the emitter does not buy its way out of its emission obligations.

Emissions trading schemes: summary
Existing ETSs have a few strengths and serious shortcomings that must be addressed. On the positive side, they all send a clear warning to investors that a carbon price is inevitable. However, until the European ETS switches to auctioned permits, it will (at best) be only moderately effective. The US RGGI has 100 per cent auctioning from the outset, but a weak initial target and hence a low initial carbon price. The details of the proposed US national ETS are still to be released.

The Australian Carbon Pollution Reduction Scheme has so many failings that it is, in my assessment, beyond redemption. It will give billions of dollars worth of free emission permits to the big greenhouse gas polluters, reinforcing their current positions. It will encourage growth in the EITE industries. It will allow unlimited overseas offsets of the unconditional emission target, thus further fostering business-as-usual within Australia. It is structured in a way that emissions reductions by households will simply reduce the emission reduction task of the big greenhouse polluting industries. Its cap on the carbon price, initially A$40 per tonne of CO_2, will ensure that, on its own, it will not allow the lower cost sources of renewable energy to compete with dirty coal.

If you think emissions trading is complicated, you are right. For carbon pricing, I recommend a carbon tax on the grounds of its simplicity, enforceability and the fact that it doesn't transfer large amounts of taxpayers' money to greenhouse-intensive industries and financial services. The only exemption to the carbon tax should be a border tax adjustment for EITE industries. In practice, even with a carbon tax, we must be vigilant. The governments of European countries with carbon taxes have granted exemptions or concessions to their major industries.

Allocate individual carbon rations

A policy that has been proposed as an alternative to both a carbon tax and an ETS is to allocate an annual ration of carbon to every person in a country. Then, when purchasing electricity, oil or gas, a 'debit card' is used to subtract the purchase from the ration. If someone's ration is exhausted, they can purchase a supplementary amount at a market.

This proposal is popular with many environmentalists because everyone participates and everyone has to take responsibility for part of their emissions. Unfortunately it has a big disadvantage: it can only work for a person's direct emissions that are easily measurable from purchasing energy: electricity, gas, petrol and air tickets. It cannot take into account the indirect emissions embodied in products and services, including food, buildings, furniture, appliances and equipment. Furthermore, it cannot take into account emissions from export industries, which are huge in Australia for example. Both a carbon tax and an ETS have the advantage compared with *carbon rationing* that, if they are applied '*upstream*', that is at the points of primary energy production (coal-mines, oil and gas fields) and import (the port of entry) rather than to end-users, the carbon price on fossil fuels then flows 'downstream'. This means it impacts on all the products made with energy and hence to end-users of energy. Individual carbon rationing is a good idea for individuals to follow voluntarily, or as a mandatory complementary measure, but it is no substitute for a carbon price.

Terminate subsidies to fossil fuels

Many countries, including highly industrialised ones, give large economic subsidies to the production and use of fossil fuels. The major parts of these subsidies are perverse, in the sense that they are economically inefficient as well as environmentally damaging.

The global energy sector receives over US$240 billion per annum in subsidies to fossil fuels. Of this coal receives at least US$53 billion per annum. This estimate includes coal's share in the subsidies for electricity generation in OECD countries only.[30] In Western Europe and Japan, subsidies to coal production in 1991–92 were equivalent to providing a domestic producer price that is more than three times the import price in Belgium, Germany and Japan, two times in Spain and 40 per cent higher in France and the UK. Furthermore, assistance per coal miner was about US$90 000 per year in Belgium and West Germany, US$38 000 per year in UK and over US$100 000 in France.[31] In the USA, Senate Energy Bills S.597 and S.14 provided coal subsidies in the 2003 fiscal year of US$4.8 billion, comprising US$2.2 billion for tax breaks and US$2.8 billion for direct subsidies.[32] This does not include research and development funding which is also measured in billions. The subsidies to the production and use of fossil fuels in Australia amount to over A$10 billion per year, and in the past have mostly gone to oil.[33] Recently, federal and state governments have provided a burst of funding for coal with CCS. As part of the economic restructuring process, these subsidies should be terminated and replaced with *temporary* subsidies to build up the renewable energy industries until they dominate the energy supply market. Actually, unlike the fossil fuel subsidies, the principal renewable energy subsidies (portfolio standards and feed-in tariffs) generally have sunset clauses.

Implement additional incentives for renewable energy

Renewable energy deserves incentives additional to the carbon price for the reasons given under fallacy 13 of chapter 2. Therefore, complementary measures are needed for stimulating the market for different renewable energy technologies. Several measures can be used either separately or in combination.

Renewable portfolio standard

When electricity generators or retailers must include a given percentage of renewable energy in their generation mix, this government requirement is called a *renewable portfolio standard* (RPS). An RPS may mandate either a percentage of renewable energy in general or a portfolio of different renewable energy sources, each with its own specified percentage contribution.

In most RPS schemes, the purchase of a new renewable energy system creates tradable certificates, each representing one megawatt-hour (MWh) of the quantity of electricity generation expected over the lifetime of the system. The monetary value of the certificate is determined by the artificial market that has been created and fluctuates over time, depending upon supply and demand for the renewable energy technologies. The bigger the renewable energy target, the higher demand for certificates and hence the higher their price. In practice, RPS is promoted as a low-cost means of subsidising renewable energy and most jurisdictions have chosen modest targets. The value of the certificates is then (at best) sufficient to bridge the price gap between conventional coal power and the least-cost renewable electricity sources, namely wind power and the cheapest forms of bioelectricity. The additional cost of RPS is spread out over the cost of grid electricity for all users.

RPS schemes are used in the USA, Canada, the UK, Sweden, Italy, Belgium, Poland, Romania, Japan, Thailand and Australia. In the latter, they are called the '(Mandatory) Renewable Energy Target'. In some places they are called 'green certificate schemes'. Common design flaws of the mechanism include lack of national coverage, weak penalties for non-compliance, inclusion of pre-existing renewable electricity (usually large hydro), major exemptions and fixed tradable certificate prices set at levels that are either too high or too low.[34]

Experience with RPS schemes is that they tend to assist only the lowest cost renewable energy technologies. Since that is what they are designed to do, it should not be interpreted as a flaw. Nevertheless, it indicates that other incentives are needed for the more expensive renewable energy sources with high potential, such as solar electricity.

In Australia, a national Mandatory Renewable Energy Target for 2010 was introduced in 2000.[35] Its weaknesses were its tiny magnitude, namely an additional 9500 GWh per year in each year from 2010 to 2020,[36] and the inclusion of some existing hydroelectricity. Its strengths were that its penalty for non-compliance, $40 per MWh, was just sufficient to bridge the price gap between wind power and dirty coal power. Overseas experience and common sense suggest that RPS works even more effectively when the penalty price is kept above the certificate price. For several years, RPS boosted the sales of hydroelectricity, wind power and solar hot water in Australia. By 2006, sufficient renewable energy had been installed to meet the tiny target, the price of renewable energy certificates collapsed and the boom in wind power became a bust. Before the November 2008 federal election, Rudd Labor promised to expand the target to an additional 45 000 GWh per year by 2020, but upon winning government delayed implementation for at least two years, as discussed in chapter 2.

Potential design flaws in the proposed new scheme[37] are the continuation of the accreditation of solar hot water, which could take a large part of the scheme to the disadvantage of wind power and bioelectricity, the exemption of EITE industries from paying the higher electricity price, and the rapid phase-out of the scheme between 2020 and 2030, which may be before the carbon price has risen sufficiently to take over from the expanded target. In addition, the rapid phase-out will undermine the prospects of renewable energy developers obtaining loans after about 2015, because project lenders generally set a 15-year debt repayment term.[38]

Feed-in tariff

A feed-in tariff (FIT) is a premium price paid for units of electricity sold to the grid by renewable energy generators who could be individual households or large power stations. 'Premium' means that the FIT is much greater than the price per unit of electricity purchased from the grid. Government mandates the price and the additional cost of FIT is spread out over the cost of grid electricity for all users. While RPS determines the quantity of renewable energy generated, FIT determines the price.

As with RPS, there are diverse types of FIT. Different tariffs can be paid for different renewable energy sources and even in some cases for different geographic regions of the same country. The successful FITs in Europe are *gross* FITs, which means that every unit of renewable electricity sold to the grid receives the high tariff. Several Australian states[39] have introduced *net* FITs, which means that the high tariff is only paid on the *difference* between renewable electricity sold to the grid and electricity purchased from the grid, provided that difference is positive. For many households, the difference is zero or negative and the scheme is of no value to them. Some jurisdictions limit FITs to the residential sector and/or limit the size of the renewable electricity installation that is eligible to receive the FIT.

FITs have been particularly successful in Germany, Spain and Denmark in encouraging the expansion of renewable energy sales and manufacturing industries. At the end of 2007, Germany had the world's largest installed wind power capacity (22 300 MW) and generated about 7 per cent of its electricity from the wind. Germany also had the world's largest solar photovoltaic capacity (3800 MW out of a world total of 7800 MW). Denmark, which also has FIT as well as other incentives for renewable electricity, generates 20 per cent of its electricity from the wind and plans to increase this contribution. Spain is generating 10 per cent of its electricity from the wind from 15 100 MW of installed wind capacity.[40]

Tax concessions

Tax concessions can be given on either the capital cost or the electricity generation of renewable energy systems. Such tax deductions can be an effective alternative incentive for an investment of high capital cost, such as a large solar power station. In the USA, the production tax credit is worth 2 c/kWh in 2009. It has given a valuable, if intermittent, support to the generation of wind power.

Discussion of complementary measures

With such a large selection of complementary measures, is there an optimal combination that we could recommend? As Germany has shown, FITs can be applied to a wide range of renewable energy sources, with lower tariffs being paid for the lower cost renewable energy sources such as wind power and landfill gas, and higher tariffs to higher cost sources such as solar power. Provided that investors are given security by legislation that FITs will not be suddenly discontinued, FITs will encourage both investment and reliable performance.

A well-designed RPS fosters low-cost renewable electricity sources. Even if the market price of renewable energy certificates is high, power station developers will tend to install the lowest cost sources first. To apply RPS to high-cost sources requires a separate target for each technology. Alternatively, a wider diversity of sources can be achieved by combining an RPS with FITs.

Both RPS and FIT are paid for by all electricity consumers, through an increased price of grid electricity. Hence both schemes are limited by the magnitude of the increase in electricity price that consumers will tolerate. However, they both aim to expand the market for renewable electricity and thus bring down its price. For this reason, governments generally decrease the value of FITs over the years according to a predetermined plan and place sunset clauses on RPS. Eventually, as renewable energy industries mature and a carbon price increases, there should be no need for either of these complementary measures.

From the published literature, I conclude that several types of complementary measures can work. We could have an RPS system with different targets for different renewable energy technologies, or FITs with different tariffs paid for different technologies, or a combination of RPS and FITs. In addition, we may need tax concessions for large renewable energy power stations.

Set regulations and standards
for energy efficiency

As mentioned in chapter 2 and discussed in more detail in *Greenhouse Solutions with Sustainable Energy*, energy efficiency has huge potential, much of it cost-effective, but its wider dissemination is inhibited by a number of market failures. Therefore, regulations and standards are essential to unleash its potential. At present, some countries have energy performance standards for new buildings and some appliances, but few countries have comprehensive programs. Mandatory energy performance standards are needed for *all* buildings, old and new, and for all new energy-using appliances and equipment. Naturally, in the case of buildings, these standards must differentiate between existing and new buildings and between geographic regions.

To accompany the standards, energy ratings are essential. Buildings should be required to have energy ratings before sale or rental, and the ratings must be disclosed prominently on sale and rental contracts.[41] Ratings and labelling are also needed for energy-using appliances and equipment. Nowadays energy star labelling exists in many countries, but the range of appliances covered is often small and so is the number of stars.

Electric resistance hot water systems and incandescent light bulbs should be phased out entirely. This process has commenced in Australia and the USA.

Foster a socially just transition

Whatever systems are chosen, governments need to consider the impact of a carbon price and mandatory energy efficiency standards on low-income earners, especially those living in private rental housing. They also need to be cognisant of the loss of jobs in the fossil fuel industries and those depending upon them. Social justice NGOs are becoming increasingly interested in these issues.

Assisting low-income earners

Carbon pricing increases energy costs to a degree that may be significant for low-income earners. A solution is for the government to provide a home visit by a trained worker, which is free of charge for low-income earners. The service would identify and implement low-cost energy efficiency measures that do not involve changes to the building envelope – walls, roof, floor and windows – and so do not conflict with the rights and responsibilities of landlords. Measures could include water-efficient taps and shower roses, compact fluorescent light bulbs, weather sealing, curtains and pelmets, refrigerator check, advice on energy efficient appliances, and insulating blankets and adjustment of thermostat for hot water tanks. These measures can be funded from a small fraction of the revenue received by government from the carbon price. This approach enables the recipients to participate in cost-saving emission-reduction solutions and so is better than purely financial compensation.[42]

Mandatory energy performance standards are essential for existing buildings, but they give an incentive to landlords to increase rents in order to cover the additional costs. Low-income tenants are particularly disadvantaged by this situation. A combination of partial financial compensation to the landlord and to the tenant may be the best solution. A step in the right direction is the Australian government's $1000 rebate to landlords for insulating tenanted housing.

Assisting displaced workers

In chapter 2, we showed that the transition to a sustainable energy system could create several times more local jobs than would be lost in the fossil fuel industries. In practice the transition will occur in discrete steps, as coal-mines and coal-fired power stations are closed. Therefore, government assistance will be required to relocate and retrain displaced workers and to provide pensions for those who cannot readily be retrained or relocated. In addition, governments should offer incentives for the manufacture of components of the new sustainable energy technologies to be located in coal-mining regions. Revenue from a carbon tax or 100 per cent auctioning of emissions permits could easily fund these programs.

International equity

The need for a process similar to Contraction and Convergence, with a large transfer of wealth from developed to developing countries, is discussed in chapter 1.

Build infrastructure

Governments should take responsibility for building infrastructure for a sustainable future. The most urgently needed items are transmission lines for renewable electricity sources such as large-scale wind farms and solar power stations, and railways for both inter-city and intra-city operation. Light rail in particular can play an important role in many cities. Cities faced with the twin challenges of climate change and peak oil also need improved facilities for cycling and walking: networks of paths, secure bicycle parking, public open spaces, and means of carrying bicycles on trains and buses.

The absence of appropriate transmission links is one of the principal barriers to the expansion of wind power capacity in the USA and Australia. The huge wind energy potential of the Midwest and Rocky Mountain states of the USA and of South Australia could only be fully tapped by building new transmission lines and upgrading existing links. This is basic infrastructure, which would be funded most efficiently by national governments, not individual wind farm developers. Also, when upgrading low-voltage *distribution lines*, planners should allow for a much larger contribution from distributed energy sources, such as cogeneration plants and small solar electric systems.

Fund research and innovation

In the past and present, the lion's share of research funding has gone to fossil fuel and nuclear industries. High-priority areas for increases in funding are:

- energy efficiency;
- solar electricity, both photovoltaics and thermal;
- off-shore wind power, including floating wind farms;
- second generation biofuels;
- biosequestration of CO_2, by planting trees and grass, and the use of biochar;
- 'smart' electricity grids;[43]
- marine power, especially wave and ocean current; and
- technologies and policies for emergency climate mitigation and adaptation.

In Australia from July 2009, the long-awaited Solar Institute is expected to commence funding research into PV and solar thermal electricity out of the Energy Innovation Fund, and there is provision in the Renewable Energy Fund of A\$15 million for second generation biofuels. It is unclear whether the other research needs will be funded.

Terminate logging of native forests

Native forests hold huge quantities of stored carbon in trees, under-growth, leaf litter and soils. A recent study has revealed that this is true, not only for tropical rainforests, but also for temperate forests.[44] Large quantities of CO_2 and other greenhouse gases are emitted into the atmosphere by logging or burning native forests. If forests are allowed to regrow, the CO_2 is gradually recaptured over decades or centuries. However, even when an area cleared of native forest is replanted as a plantation, the net emissions are positive and large.

International negotiations are in progress to reduce deforestation and degradation in tropical rainforests (REDD). These should be extended to temperate forests. For all native forests, it is essential to go beyond REDD to secure permanent protection, ecologically and legally, with properly funded management.[45]

The time is long overdue for bans on the logging of native forests in developed countries, which should also impose bans on imports of timber from forests that are not rigorously certified as ecologically sustainable. If necessary, this issue has to be fought in the World Trade Organisation (WTO). Unfortunately the Forest Stewardship Council, the leading international non-profit organisation that certifies wood products, has come under increasingly severe criticism by a variety of environmental NGOs. They claim that it is allegedly relaxing its standards and supporting the logging of old-growth native forests.[46]

Cut agricultural emissions and increase biosequestration

Agriculture is a large contributor to global greenhouse gas emissions. In Australia, it is responsible for 16 per cent of emissions and is the second largest emission source after stationary energy. Energy analyst Hugh Saddler and environmental scientist Helen King present a strong case that agricultural emissions cannot be handled within an emissions trading scheme,[47] because:

- the fact that they cannot be measured accurately could undermine confidence in the whole scheme;
- most farmers cannot readily control them and so the price mechanism would simply become a tax on production without necessarily reducing emissions;
- most farmers are individually small emitters, although their collective emissions are large, and so an emissions trading scheme designed to allocate permits to a few large emitters would be inappropriate.

Therefore, alternative abatement policies for agriculture are necessary. Saddler and King suggest a carrot and stick approach to encourage a wide range of abatement options: levy and incentive payments; accreditation standards; and voluntary markets.

Agriculture also has large potential for biosequestration, that is, capturing CO_2 by plants via photosynthesis and storing the captured CO_2 in the plants, leaf litter and soil. Given the right incentives, farmers could do this by planting trees, grasses and other perennial crops;[48] by producing biochar (wood charcoal) and mixing it with the soil; and by choosing crop species that produce the maximum numbers of phytoliths (see chapter 3). More policy development is needed for agriculture.

Train the workforce

The transition to a sustainable energy future will create demand for a multitude of new professional and technical jobs and for major upgrades of knowledge and skills in many existing jobs, such as plumbing and electrical trades. A well-trained workforce is essential for the expansion of energy efficiency, renewable energy, improved urban planning, public transport and new approaches to agriculture. Given government policies to encourage deep cuts in emissions, some of the important jobs and skills will include:

- installers of insulation, solar hot water and solar PV for residential and commercial buildings;

- energy auditors and engineering consultants for energy efficiency;
- architects and builders who are competent in passive solar design and ecologically sustainable building processes;
- urban planners who understand transport planning and vice versa;
- metal-workers, electrical workers and information technology workers who can build the components of wind and solar power plants;
- electrical engineers experienced in renewable electricity generation and in power transmission;
- biological scientists and chemical engineers to research and produce biofuels;
- engineers and scientists competent in life-cycle assessment of energy and other systems; and
- facilities managers and heating, ventilation and air conditioning operators of commercial buildings.

Policy responses for training the workforce must address:[49]
- clear, strong, greenhouse reduction policy, to establish confidence in the short- and long-term direction of change;
- skills policy, including more quantity, better quality and wider scope of sustainability education in all formal spheres – primary, secondary, tertiary, vocational and technical – and also in 'skill ecosystems' involving stakeholders from business, trade unions and government agencies;
- industry policy, which should include improving the links between education/training and commercialisation, in order to move innovation into the marketplace;
- the creation of a database on green skills and workforce capabilities; and
- performance assessment and accreditation.

Stabilise population growth

As discussed in chapter 2, population growth is one of the three driving forces of increases in greenhouse gas emissions. In countries with very high per capita emissions, such as the USA and Australia, every additional person is on average responsible for much greater emissions over his or her lifetime than in a less developed country. Therefore, population growth must be ended as a matter of priority in countries with high per capita emissions and, in the longer term, in all countries.

Policy responses are straightforward in developed countries: removal of government propaganda[50] and financial incentives to encourage births; reduction in immigration quotas. Incidentally, for those who would like to increase the refugee intake of developed countries, this can be readily done in countries such as Australia where refugees form only a small proportion of immigrants. It simply means reducing quotas for professional and business immigrants, who are generally the majority of immigrants. Once a situation of Contraction and Convergence of greenhouse gas emissions has been achieved, then immigration quotas can be removed. However, policies to maintain the birth rate at replacement level will still be needed.

In less developed countries, the main policy needs for reducing births are the empowerment and education of women, the provision of contraception and the right to use it, and economic development that provides social security for the aged.[51] To assist in achieving these goals, developed countries should increase their overseas aid budgets and loosen the strings that tie them to their own businesses.

The principal forces resisting population control policies are the Roman Catholic Church, which still opposes birth control, the property and housing industry, and indeed capitalism in general, which pushes for an excess of labour in order to keep wages down. Fearful of being unjustly accused of racism or being anti-refugees, much of the environmental movement has avoided the population issue or cloaked it in generalities.[52] This is a serious campaign gap.

Create a socially just,
steady-state economy

The dogma that endless economic growth is possible and desirable is one of the principal drivers of the growth in greenhouse gas emissions, as discussed in chapter 2. The power-holders who are tied to this view push for more and more motor vehicles, power stations, mining, manufacturing, land clearing and chemicals. They exaggerate the benefits of this pathway and play down its adverse effects and limitations. The growth dogma, together with many other socially and environmentally destructive practices, is given public justification by neoclassical economic theory.

A large body of research and writing, by economists, scientists, mathematicians and others, argues for a radical change in the dominant economic system, for excellent reasons that go beyond the growth in greenhouse gas emissions.[53] This work makes the case that the dominant economic system is destructive to the majority of people as well as to the environment. It is impoverishing the majority of poor countries and transferring wealth from the poor to the rich countries. Within poor countries, it is making the poor poorer and the rich richer. It does this by imposing structural adjustment policies that:

- encourage the replacement of food crops with cash crops;
- drive peasant farmers from the land to change them into cheap labour and fringe-dwellers in the cities; and
- dismantle government programs providing education, health care and in some cases social security to the people.

This in turn drives desperately poor people to environmental destruction, such as the logging of native forests, a big greenhouse gas emitter, or fishing with gelignite.

Within many countries, both rich and poor, the economic system and the associated financial system are taking resources and wealth from the majority, and concentrating it in the hands of small wealthy elites. They do this by:

- privatising public utilities;
- centralising banking;
- giving free rein to a financial industry based on speculation with bailouts by the taxpayer; and
- transferring enormous powers to corporations and unelected tribunals such as the WTO that are biased towards corporations.

The body of scholarship[54] exposing these injustices and beginning the process of developing solutions, has been ignored by economic power-holders and the politicians and officials who follow the neoclassical economic ideology, despite the dismal economic performance of the system. The fundamental problems lie in both the economic and financial practices followed and the theory used to justify them.

Neoclassical economic theory, while mathematically elegant in some respects, is based on a series of misconceptions about the real world. For example, its followers assume incorrectly that:

- People are competing to maximise their welfare based on the purchase of products and services. They do not cooperate with one another or form institutions for mutual benefit.

- The natural environment is an infinite resource base and an infinite waste dump.

- Gross domestic product is a good measure of wealth and quality of life.

- All important transactions between people occur in markets which are, to first approximation, competitive: that is, supply balances demand and none of the buyers or sellers can influence price.

- Furthermore, the optimal use of resources – that is, when production is maximised and prices are minimised – occurs when supply and demand balance in the market.

- Economic agents who participate in markets – that is, individuals, households and firms – have complete knowledge about prices paid for any item in the market by all other agents.

- The sizes of corporations are limited, based on the notion that profits decrease or remain constant as size increases.[55]
- The best monetary system is the one that is currently dominant, which drives markets by speculation and manipulation, rather than by serving the interests of productive enterprises.

As discussed in the references cited above, all these assumptions are either completely wrong or are only valid under special conditions such as village markets. On the basis of the huge gap between economic theory and observations of society, scientist Geoff Davies argues convincingly that neoclassical economics is a pseudo-science – that it has sacrificed reality for mathematical elegance. The emperor has no clothes.[56]

It's astonishing that a theory that fails so badly to describe reality can be used to 'justify' practical policies. For example, the theory of comparative advantage is only valid under conditions that no longer apply: that is, when capital cannot be transferred freely across international boundaries. Yet it is used widely by some economists and many politicians to 'justify' policies that constrain some poor countries (and even some rich countries such as Australia) to being quarries for the rich countries. Therefore, neoclassical economics is not only a pseudo-science, but it is also an ideology which true believers and opportunist unbelievers use to manipulate and exploit the majority of people.

The solution to this situation is radical reform to both the theory and practice of economics. Considering the huge power structure that has evolved to protect this unreal, unjust and destructive system, reform is likely to be a difficult long-term process. Meanwhile, climate movement organisations (CMOs) and other NGOs can pursue the following policy actions in order to contain the problem:

- Resist moves by governments to strengthen the power of corporations and the WTO and to impose free trade agreements. This resistance is already happening, but needs support from CMOs.
- Resist bail-outs by governments of financial institutions that

face bankruptcy as a result of poor lending practices. It is far better for government to let them collapse, to provide assistance packages for the unemployed and to stimulate the market for new jobs in energy efficiency, renewable energy, public transport and other socially useful areas.

- Increase the fractional reserve[57] that financial institutions must maintain for lending to 50 per cent, in order to put a brake on speculative lending.

- Further reduce speculative trading by introducing a transaction tax, known as the Tobin tax, on all foreign exchange dealings. The tax would be a small percentage (say 0.1 per cent) of the amount transferred and the funds raised could have the additional benefit of funding a worthy cause, such as the United Nations or reducing world poverty.

- Strengthen laws and economic instruments to make polluters pay.

- Strengthen laws and economic instruments to increase the reuse and recycling of materials.

- Introduce alternatives to GDP as official indicators of economic performance, for instance the Index of Sustainable Economic Welfare or the Genuine Progress Indicator.[58]

- Explore alternative monetary systems that are stable and self-regulating and encourage local economic development.[59]

With these government policies to shape the market and restructure the economy, business and industry will adjust and reduce their direct emissions and those from the products they sell. A necessary policy is to strengthen the existing guidelines to financial institutions to discourage lending for environmentally destructive projects, such as those that are greenhouse intensive. In theory, the risk of increasing carbon prices should be sufficient restraint, but in practice governments often provide guarantees, backed up by the taxpayer or the electricity user, to developers of megaprojects against financial risk. Such guarantees should be banned. Furthermore, laws should be tightened to increase penalties for company directors for making risky speculative investments and to mandate that corporations disclose annually key environmental performance and risk indicators for their internal operations, based on a framework combining principles-based and specific reporting requirements, such as the Global Reporting Initiative.[60] Banks and other financial institutions should be required to disclose details of the aggregate environmental and social performance of their investment portfolios. Similarly, superannuation funds and other pooled investment funds should be required to disclose annually a list of their investment holdings.[61]

Policies for local government

The foregoing policies are primarily for national and state governments and to some extent for municipal governments that are responsible for whole cities. Local governments acting together can potentially make a large contribution to the reduction in national and state greenhouse gas emissions. They deserve a detailed treatment, but since space limitations do not permit that in this book, I simply list some of the issues:

- Tighten energy efficiency and solar hot water requirements in development approvals for new buildings until national or state governments set the standards.
- Plan local roads, local travel destinations, parking, footpaths and bicycle paths, in order to reduce car use.
- Waive planning approvals and other restrictions on solar hot water, solar electricity and solar clothes drying (clotheslines).
- Implement a home energy audit and implementation scheme until this is done by a higher level of government.
- Hold public training workshops on energy efficiency in the home.
- Support local community projects to install wind and solar farms.

- Make local government's own operations more energy efficient, addressing its buildings, vehicles, street-lighting, swimming pool heating, solid wastes, etc.
- Join the Cities for Climate Protection program of ICLEI – Local Governments for Sustainability.[62]

Conclusion

No single policy measure can solve the climate crisis. Instead we need a wide range of policies that span carbon pricing, regulations and standards, education and training, research and innovation, and institutional change. Policies must address stationary energy, transport and urban form, agriculture, forestry and industrial emissions other than those resulting from energy use. They must greatly reduce the emission of all greenhouse gases. They must change technology and economic structure, and stop population growth. The policies listed in this chapter, if implemented, will take us a long way down the pathway to a solution.

STRATEGIES FOR DEFEATING THE GREENHOUSE MAFIA

Imagine the climate action movement is playing a modified game of chess with the Greenhouse Mafia (GM). The stakes are the world, including most human civilisation. To complicate matters, each side has different pieces. GM has a few very powerful pieces, while the climate action movement has a wide variety of less powerful ones, knights, bishops, scholars and countless pawns.

GM starts aggressively and captures several queens, the most powerful pieces of all, which have been standing on the sidelines. However, the result is not a foregone conclusion: the climate action movement is still in the early stages of mobilising its forces and developing its strategies. To make the game even more challenging, the world is not a passive prize, waiting breathlessly to be swept away by the victor. She holds the timer for the moves, and she is becoming impatient ...

Building an organisation and developing a winning strategy for it are two aspects of the same process. Without an organisation there can be no strategy and without a strategy the organisation is useless. In this book, I define an organisation as a group with some degree of structure. For individual climate action groups, the degree of structure may vary from loose and informal to tightly defined roles. Our climate action movement as a whole, on a national or global scale, also needs some structure to be effective. That structure may involve coalitions, alliances and/or networks, which are discussed below.

In a game of chess, if two players are unskilled, the match can be won without strategy. But if one competitor has strategy, and the other doesn't, there are no prizes for guessing which one will win in the end. Strategy is the planning and conduct of long-term campaigns to achieve broad goals. Nonviolence expert Gene Sharp offers a similar definition, describing strategy as 'charting the course of action which makes it most likely to get from the present to a desired situation in the future'.[1] In its book *Organizing for Social Change*, the Midwest Academy, a training institute for progressive social change, defines strategy much more specifically as 'an approach to make a government or corporate official do something in the public interest that he or she does not otherwise wish to do'.[2] This definition deliberately targets one or more individual decision-makers rather than an organisation. It states that strategy requires your organisation to exercise power against that individual. As I see it, this offers a valuable perspective, discussed below, but is not sufficiently broad for all the needs of social change movements.

Tactics are the individual steps or tools used in carrying out a strategy. Tactics are limited, short-term courses of action on the long-term strategic pathway. They may involve face-to-face lobbying, media events, an educational campaign, a lawsuit, and sit-ins, for example. We discuss tactics in chapter 6. In the present chapter we explore the fundamental concepts of organisation, strategies and power.

The power of organisation

In the face of the enormous wealth and political influence of the vested interests in greenhouse gas pollution, the climate action movement has three principal strategic strengths:

- It serves, with integrity, humankind as a whole, instead of the owners and shareholders of destructive industries. This provides it with moral and credibility advantages.
- It is sufficiently diverse to influence groups and individuals in almost all walks of life.
- It can potentially call upon the support of far greater numbers of people to carry out its campaigns.

The ethical/moral issues are touched on in chapter 1. The present and following chapters are devoted to using the strategic and tactical advantages of the numerical superiority and diversity of the climate action movement.

In several countries, public opinion polls show that the majority of the population rates global warming as one of the most serious and urgent issues faced by society in the 21st century. For instance, an exit poll conducted in marginal seats at the November 2007 Australian federal election found that climate change was the third most important issue distinguishing the major parties. Of those in marginal seats who voted Labor into power, 85 per cent said they wanted Australia to be a leader in international action and negotiations and 85 per cent wanted Australia to set greenhouse gas reduction targets of 20 per cent by 2020 and 80 per cent by 2050.[3]

Subsequent public opinion polls found that public concern about climate change fell marginally in the six-month period since March 2008, but remained high, down from 89 per cent to 82 per cent, despite the financial crisis.[4] So a large majority is already sympathetic to climate action. What is lacking is the mobilisation and organisation of the majority into a large movement that exerts irresistible pressure on national and state governments and business.

Saul Alinsky, a professional organiser, who has worked for decades to assist the 'have-nots' of the USA, writes that:

> ... every move revolves around one central point: how many recruits will this [action] bring into the organisation, whether by means of local organisations, churches, services groups, labor unions, corner gangs, or as individuals. The only issue is, how will this increase the strength of the organisation ...
>
> Change comes from power, and power comes from organisation. In order to act, people must get together ... Power and organisation are one and the same.[5]

Some of you may feel uncomfortable talking about power and targeting it against individual decision-makers. 'Surely,' you may say, 'all we have to do is give the decision-makers the facts and they will make the right decisions.' Unfortunately that is not the way of the world.

For example, Australia's new Labor Government keeps a close watch on public opinion polls and is aware that the public wants genuine, substantial climate action. Yet it has chosen deliberately to delay the actions that it promised before the November 2007 election that brought it to power. Several of those promised actions have been delayed for years. Meanwhile, it is implementing only token actions that will not significantly change the energy supply system for the foreseeable future (see chapters 2 and 4). Furthermore, it has refused to stop the logging of native forests, another big source of emissions. Clearly the government has made the political decision that the power of the big greenhouse gas polluters is greater than the power of the much larger numbers of unorganised masses that elected it. No doubt the government has also considered that, in a two-party electoral system, it only has to appear to have policies that are slightly better than the Opposition's to be re-elected. Only by changing the balance of power can the climate action movement change government policy.

Some writers distinguish the distributed power of a mass movement from the concentrated power of a government or a very large industry or corporation, arguing that distributed power is a better form of power than concentrated power. They may interpret 'absolute power' as 'concentrated power' in Lord Acton's famous quotation 'Power tends to corrupt, and absolute power corrupts absolutely'.

It is true that, when hundreds or thousands of climate movement organisations (CMOs) are each acting locally around the country and around the world, they are exercising a distributed form of power. But, the power of vested interests is held centrally, in states, nations and international agreements, and so I have to agree with Alinsky that there are times when a movement must organise and cooperate on a large scale and focus its own power into a concentrated form, just as a lens or mirror focuses solar energy for beneficial purposes. If we had a century to stop runaway global warming, the local actions of hundreds of CMOs in each country *might* be sufficient to transform government policies in state, national and international spheres. Unfortunately, in the case of global warming, time is of the essence. The climate action movement must build its organisational capacity to conduct operations from time to time on larger scales than the local.

The opening sentence of Gene Sharp's classic book, *The Politics of Nonviolent Action*, reinforces this: 'Some conflicts do not yield to compromise and can be resolved only through struggle.'[6] He goes on to write (p 7):

Unlike utopians, advocates of nonviolent action do not seek to 'control' power by rejecting it or abolishing it. Instead, they recognise that power is inherent in practically all social and political relationships ... They also see that it is necessary to wield power in order to control the power of threatening political groups or regimes.

Sharp's view is based on a detailed study of nonviolent actions over millennia from around the world. He states (p 8) that, although governments appear to exercise 'monolithic' power and that the people appear to be 'dependent upon the good will, decisions and the support of the government or any other hierarchical system to which they belong', reality is the opposite. Drawing upon many historical examples, Sharp argues that governments actually 'depend on people, that power is pluralistic, and that political power is fragile because it depends on many groups for reinforcement of its power sources'. The power of government and corporations depends upon the consent of the people – and we the people can together withdraw that consent. In this context, Sharp is referring to a much wider range of actions than voting at elections.

A large and growing body of experience exists in nonviolent social movements in fields of social justice, environmental protection, consumer rights, community development and peace. Some of the leading groups and individuals have been recognised by the Right Livelihood Awards,[7] which are sometimes described as 'the alternative Nobel Prize'. The accounts of these successes are inspiring and I can recommend them as an effective tonic when the power of the Greenhouse Mafia seems overwhelming.

People are highly motivated to withdraw consent when they are suffering invasion or a repressive government, the situations considered by Sharp. They will also struggle for better conditions if they are underpaid and insecure workers, or people living in slums or in a highly polluted area, such as the infamous Love Canal[8] in the USA. In such cases, people have a clear direct interest in improving their conditions and the social movement's main challenge is to convince them that they can achieve change by working together.

In the case of climate change, the majority of people do not yet see immediate strong direct impacts on themselves. Possible exceptions are some people who experienced the extraordinary heatwave in Europe in 2003, or the residents of New Orleans whose homes were destroyed by Hurricane Katrina in the USA in 2005, or farmers experiencing drought in the Murray-Darling Basin of Australia. Here is fertile ground for CMOs. However, for the majority of the population, these events have occurred 'out there' and are not yet of sufficient direct relevance to move them beyond concern to action. If they are living in coastal cities, their time will come. Meanwhile, one of the principal challenges for a CMO and its organiser is to jolt the people out of their complacency. To do this, requires some organisation and the development of a well thought-out strategy, starting with a collective choice of broad and specific goals.

Developing goals

First, your CMO (and the climate action movement as a whole) must develop a shared vision of the preferred future and a set of broad goals. Depending on the type of CMO, the vision and broad goals could be global, national or local. In all cases they will comprise a broad conception that the vast majority of group members accept. For example, in fighting global warming, a vision of an international coalition or alliance could be the achievement of Contraction and Convergence, a process by which all countries reach the same safe level of average per capita greenhouse gas emissions by (say) 2050. On the national level, it could be a series of national greenhouse gas emission targets from 2015 to 2050. On a local level, your vision could be that your region will emit, let's say, 30 per cent less greenhouse gases than the 1990 level by 2020.

The next step is to break down your vision into several bite-sized chunks that are achievable within a reasonable timeframe. Some activist groups call these specific goals 'issues'. A national issue could be the implementation of a gross feed-in tariff for renewable energy by a specified date. A state issue in the USA or Australia could be the defeat of a proposal to build a new conventional coal-fired power station (which is a national issue in the UK). Local issues could be:

- winning over the local government and local member of state parliament to publicly support the CMO's local greenhouse target;

- convincing the three biggest industries in the region to cut their emissions by 30 per cent;

- building a community-owned wind farm or solar power station.

Most CMOs will choose several issues, some of which can be achieved within a few months, in order to post some wins on the board, and one or two more difficult, long-term issues, to stretch the group.

Issues should be evaluated carefully in a participatory process. According to activist groups, they should satisfy most of the following criteria[9] that the issue:

1 Improves people's lives

This is easy to demonstrate for (say) implementing a residential energy efficiency program for low-income housing or building a new railway. However, for setting greenhouse targets, the CMO will have to broaden the horizons of its members and the public to think of the long-term benefits.

2 Builds people's confidence in their own power

When the goal is reached, members of the CMO should feel that they achieved it, not outside experts or politicians.

3 Strengthens the power of the people in relation to the decision-makers

This includes internal strengthening, such as building up membership of the CMO, as well as external outcomes, such as new infrastructure, laws, regulations and standards, institutions and economic instruments.

4   Is winnable

This is especially important for the first few issues tackled.

5   Is widely and deeply supported in the community

This is essential for gaining the support of large numbers of people and for presenting a unified message to decision-makers.

6   Can be presented clearly and simply to the media and public

This tactic, called *framing* the issue, is an art-form in itself and whole books have been written about it.[10] It is both part of the strategy of choosing the issue and part of the tactics and language of communicating it. We discuss it further in chapter 6.

7   Has one or more particular individual decision-makers as a clear target

According to the view of several activist groups, including the Midwest Academy, without an individual target, the issue is likely to be too broad and diffuse to be achievable. However, I think that in some cases of particularly egregious behaviour by a corporation, it can be fruitful to target the whole corporation, especially when the CEO is unknown to the public or there is a turnover of CEOs. Successful examples were Greenpeace's campaign against Shell to stop its proposed disposal of the Brent Spar oil platform by sinking it in the North Atlantic,[11] the campaign by the citizens of Minamata in Japan to identify the cause of the crippling disease that afflicted them,[12] and the campaign to obtain compensation for asbestos victims from the company James Hardie.[13]

8   Has a clear time-frame convenient for the CMO and compatible with external events

Key dates that you should take into account are parliamentary sittings, elections and the release of the government's budget.

9 Does not split your principal supporters, actual or potential

A possible example is, for a local group supporting the construction of a wind farm in their region, to ensure that the whole community benefits financially, not just the properties upon which the wind turbines are erected. Your CMO could do this by lobbying the local government to require the developer to make a financial contribution for a community facility.

10 Builds knowledge and leadership within the CMO

Your CMO should construct the campaign to offer many roles to group members and a mentoring process for less experienced campaigners.

11 Is consistent with the CMO's values and vision

For example, if your group claims to be motivated by concern for future generations, it must surely also support policies to assist just transitions for coal miners and low-income earners who will be hardest hit by increasing energy prices resulting from carbon pricing.

Each goal or issue requires its own strategy or campaign plan.

Developing strategy: SWOT analysis

A useful initial exercise for assessing a specific goal or issue is a SWOT analysis, which is a strategic planning method to identify the Strengths, Weaknesses, Opportunities and Threats/barriers for the goal. Strengths and Weaknesses involve the group's internal resources in relation to the goal. Opportunities and Threats involve conditions external to the group that may influence its ability to reach its goal. The results of the SWOT analysis may indicate that one of more of your goals is unattainable. Thinking positively, SWOT analysis should lead your group to ask:

- How can we use each Strength?
- How can we address each Weakness?

- How can we seize each Opportunity?
- How can we avoid, diminish or counter each Threat?

If your goal is to convince the three biggest industries in your region to cut their emissions by 30 per cent by a given date, then group Strengths might be that a business member of the group plays golf with the CEO of one company. This Strength can be used to gain introductions – however it is unlikely on its own to achieve the outcome you seek. A Weakness may be a lack of knowledge of the details of industrial processes in the region. This could be overcome by direct enquiry of the industry and/or seeking advice from appropriate academics and businesspeople. Opportunities may arise from the visit of an overseas expert, or the publication of a prestigious report that supports one of your goals, or the occurrence of a severe cyclone/hurricane at a higher latitude than usual. You can seize upon these opportunities to draw industry CEOs into a workshop and members of the public into a community meeting. A Threat, for example that the government proposes to give free emission permits to one of the big local industries, could be diminished either by media publicity or by forming an alliance with the other industries that are not eligible for free permits. You could address a more generic Threat, such as an unsympathetic editor of the local newspaper, by discussions or, failing that, a program to build constructive relationships with other media. For a small, simple campaign proposal on an easily achievable issue, SWOT analysis may be all that is needed to assess it.[14]

Developing strategy: Midwest Academy Strategy Chart

For much greater depth and breadth than SWOT analysis, your CMO could use the Midwest Academy Strategy Chart.[15] This chart is specifically designed for campaigns aimed at winning something from someone. It is not designed for election campaigns or campaigns that are primarily educational or for dealing with internal problems within your own organisation. The chart has five columns, one for each of the principal strategic elements:

1 Long-term, intermediate and short-term goals.
2 Organisational considerations.
3 Constituents, allies and opponents.
4 Targets.
5 Tactics.

Members of your CMO can work together to fill in the chart, proceeding from the first to the fifth column. It should be noted that, if the group changes an element in one column, corresponding changes have to be made in the other columns. Thus the chart is more like a spreadsheet with qualitative data than a table.

The Midwest Academy Manual applies the chart in detail to a hypothetical campaign to win tax reform in an unnamed US state. In table 5.1, I have partially filled in the chart for the case of a campaign by a state-based CMO, comprising an alliance of climate action groups, to stop a proposal for a new conventional coal-fired power station, a substantial goal.

Columns 1–3 are largely self-explanatory. In column 4 a secondary target is someone who has more power over the primary target than the CMO does, while the CMO has more power over the secondary target than it has over the primary target. In the example, possible secondary targets are the CEOs of financial institutions that might be considering making loans to the developer of the power station. Financial institutions are becoming anxious about the increased risks of lending for such a purpose in the face of new or increasing carbon prices and they dislike adverse publicity for loans to environmentally damaging projects. Other possible secondary targets could be donors to the government who would be disadvantaged by the power station.

Column 5 is always filled in last, since tactics must be part of the campaign strategy and not an end in themselves. The Midwest Academy points out that for every tactic it is worth considering:

• someone who does it;

TABLE 5.1 Midwest Academy Strategy Chart applied to stopping coal power

Goals	Organisational considerations	
1. Long-term goals • Stop construction of the proposed power station. **2. Intermediate goals** • Develop and publish an alternative energy scenario. **3. Short-term goals** • Have a question asked in parliament: 'Is the power station necessary and why were alternatives not considered?'	**1. Resources to put in** • Salary for campaign coordinator, halftime over 12 months: $40k. On hand $30k; to raise $10k. • Consultant on alternative energy scenario: to raise $20k. • Volunteer campaigners to be chosen: 6 part-time over 12 months: total value $100k full-time equivalent. • Free use of part of our CMO's office space, desks and computers for 12 months: value $5k. • Stationery, phone, printing of report, postage: to raise $5k. • Media consultant: to raise $4k.	**2. What we want to get out of it** • Cover all dollar expenses by contributions from alliance members and campaign fundraising. • Gain new organisational members for the alliance, most likely from local wine-growing and horse-breeding industries in the valley and the national insurance and reinsurance industry. • Gain at least 100 new individual members for affiliated climate action groups. • Gain and train 6 active volunteer part-time campaigners. **3. Problems to solve** • Climate action groups are divided on whether natural gas should be included in the alternative energy scenario.

• someone to whom it is done;

• a reason why the person to whom it is done doesn't want it done and will make a concession to you if you stop doing it.

While this may be generally true, my activist colleagues and I have found that some tactics are still worthwhile if they empower members without doing something to someone.

Coalitions, alliances and networks

The power of a social movement can be increased greatly by building strategic coalitions, alliances and networks. Although these terms are

Strategies for defeating the Greenhouse Mafia

Constituents, allies and opponents	Targets	Tactics
1. **Constituents and allies** • Electrical trades union • Metalworkers union • Education unions • Renewable energy industry association • Environmental business association • Insurance industry • Local wine companies • Local horse-breeders • Student organisations • Some local governments • Some individual coal-miners 2. **Opponents** • Coal industry • Trade union for coal-miners • Chamber of Commerce • Electricity Generators Association • Resource industry lobby groups, think-tanks and front organisations	1. **Primary targets** • Head of state government • State energy minister or equivalent • Government Standing Committee on Energy 2. **Secondary targets** • CEOs of financial institutions • Donors to the government who would be disadvantaged by power station	(Only categories of tactics are listed here. Specific tactics that could possibly be used are discussed in chapter 6.) • Lobbying power-holders • Nonviolent direct confrontation • Educational activities • Media coverage, questioning the need for the power station and promoting our alternative scenario • Legal action • Setting up alternatives

SOURCE The template for this table is © Midwest Academy, 27 East Monroe, 11th Fl, Chicago, IL 60603, USA; <http://www.midwestacademy.com>.

NOTES I thank Midwest Academy for permission to use its Strategy Chart template, which I have applied as an example to a particular issue. For a detailed discussion of the Midwest Academy Strategy Chart and a different, more focused example, see Bobo et al (2001) listed in 'Key readings and websites'. My application of the chart is not intended to be comprehensive – the reader may think of additional entries within the given framework.

used flexibly, it is helpful to make some distinctions between them. A coalition usually involves a long-term relationship among its member groups. It may have a formal structure with staff. An alliance usually involves a shorter-term relationship among groups that is based on a single issue or objective. It may have temporary staff or none at all. A network is generally a loose flexible association of groups and sometimes individuals, brought together for sharing information and ideas. It rarely has staff dedicated to the network. Academic and environmental activist Timothy Doyle summarises the advantages of networks:

> Environment networks are primarily created to collate and disseminate information ... First, it helps to avoid duplication of effort. Second, when correctly digested, information is knowledge. With access to certain knowledge, environmentalists are less inclined to 'reinvent the wheel', and campaign strategies can be created with this collective knowledge. Finally, access to information provides groups with a sense of belonging to something far larger than themselves: a politically powerful social movement. This feeling inspires greater effort and achievement.[16]

Doyle's book goes on to give some valuable insights into the politics and power relationships in environmental networks.

Coalitions and alliances with influential organisations

CMOs may participate in coalitions and alliances for specific purposes with influential organisations, such as large businesses, trade unions and professional organisations. Developing such relationships usually involves protracted negotiations to build mutual understanding, even trust, and to identify potential win-win cooperation. As in the case of lobbying, CMOs must bring their power and influence to the table. For instance, they can support trade unions in their requests for a just transition to a sustainable future for workers in the coal industry. In return they could receive support for policies to develop and promote renewable energy.

The advantages of coalitions and alliances are that they can make weighty public statements of common purpose and policy, fund important research and consulting reports, fund projects to construct alternative technology projects, and form joint delegations to lobby power-holders. Alliances can also split powerful blocs of greenhouse deniers. These actions can influence both the principal power-holders and the community at large. Examples of some alliances are given in box 5.1.

Examples of alliances between CMOs and other organisations

Ceres is a large US coalition of investors, environmental groups and other public interest groups working with companies to address sustainability challenges such as global climate change. In 2008 Ceres joined with five US corporations – Levi Strauss & Co, Nike, Starbucks, Sun Microsystems and The Timberland Company – to announce the launch of a new business coalition, Business for Innovative Climate and Energy Policy, calling for strong US climate and energy legislation in early 2009. The coalition's key principles include stimulating renewable energy, promoting energy efficiency and green jobs, requiring 100 per cent auction of carbon allowances, and limiting new coal-fired power plants to those that capture and store carbon emissions.[17]

The Australian Business Roundtable on Climate Change is a coalition of six large businesses – BP Australia, Insurance Australia Group, Origin Energy, Swiss Re, Visy Industries and Westpac – convened by the Australian Conservation Foundation. The roundtable commissioned the CSIRO to determine climate impacts on Australia and the Allen Consulting Group to model the economic effects of producing a 60 per cent reduction on year 2000 emissions by 2050, for its report *The Business Case for Early Action*. It is an ad hoc group that meets for such specific tasks.[18]

The Central Victorian Greenhouse Alliance brings together 14 local governments and nine businesses and other organisations into joint projects. Its mission, from the beginning, has been to lead a thriving region to a 30 per cent reduction in greenhouse gas emissions by 2010 and to zero net greenhouse gas emissions by 2020.[19]

The Southern Cross Climate Coalition is an alliance between the

> Australian Conservation Foundation, the Australian Council of Social Service, the Australian Council of Trade Unions and the Climate Institute. Its first statement, released in July 2008, called for a broadly-based emissions trading scheme with a tight cap on emissions and 100 per cent auctioning of emission permits. It also supports measures to protect low-income households, a mandatory national energy efficiency target, the promised Mandatory Renewable Energy Target of 20 per cent of electricity by 2020, and a 'green skills' training program. However, it does not specify a short-term greenhouse target for Australia, or recommend an initial carbon price, or support a ban on new dirty coal-fired power stations.[20]
>
> More examples are given in the appendices.

Coalitions and alliances also have disadvantages and risks, most notably the danger of some CMOs giving support to the preferred strategies of the power-holders, thereby splitting the movement. Some CMOs have formed alliances with peak industry organisations that support the development of coal power with carbon capture and sequestration (CCS). An example is the alliance formed by the environmental NGOs WWF-Australia and the Climate Institute with the Australian Coal Association and the Construction, Forestry, Mining and Energy Union to lobby for a national CCS strategy.[21] Unfortunately the alliance did not call for equal or greater support for renewable energy.

An example that demonstrates even greater risk for CMOs is the United States Climate Action Partnership (USCAP), which is a group of businesses and leading environmental organisations 'committed to a pathway that will slow, stop and reverse the growth of US emissions while expanding the US economy'.[22] Not only does USCAP support CCS, but it also endorses the construction of new coal-fired power stations without CCS.[23] It is disappointing that NGOs such as the Natural Resources Defense Council, the Pew Center on Global Climate Change, the Nature Conservancy, Environmental Defense and the National Wildlife Federation have joined this alliance. Surely all genuine CMOs should oppose all proposals for new coal-fired power stations that do not have CCS at the outset. A promise that a new power station will be 'CCS-ready' is worthless in the context of the current poor status of the technology and rapid global climate change.

Alliances and networks among CMOs

At this point I should remind you that I'm using 'climate movement organisation' (CMO) to describe all organisations that are partially or totally devoted to deep cuts in greenhouse gas emissions. Within this definition, CMO includes NGOs such as Greenpeace and WWF, that work on a wide range of environmental issues in addition to the climate crisis, as well as groups such as Clean Energy for Eternity, which are 100 per cent dedicated to climate action. Within the climate action movement alliances and networks are needed to coordinate campaigns in state/provincial, national and international spheres.

A social movement is likely to be more effective if it fosters links between a diverse range of CMOs with different ideologies, strategies and tactics, provided they all follow a nonviolent pathway. An alliance of environmental NGOs, professional, student, academic, trade union, business, local and faith groups can potentially appeal to a large majority of the wider community. This is another a clear advantage of the climate movement over the vested interests opposing greenhouse gas reductions. For example, the Apollo Alliance in the USA is an impressive coalition (more than an alliance) of environmental NGOs, trade unions, renewable energy businesses and other groups.[24]

Diversity also impacts on power-holders. For instance, nonviolent direct actions by radical groups of the climate action movement can encourage power-holders to negotiate with less radical groups. Thus the climate action movement can follow a multi-strategy where:

> revolutionary and moderate groups ... can operate as two wings of a common campaign. Here, radicals demand attention and foment controversy ... In contrast, reformists offer a sane, rational alternative ... Governments fear the extremists and meet with the reform-

ists. Strategic divisions within the movement can become, briefly, a kind of political resource.[25]

Nina Hall and Ros Taplin, then environmental researchers at Macquarie University in Australia, point out that 'This observation may be equally valid for other proposed dichotomies, such as 'insider versus outsider', or 'confrontational versus reformist'.[26] Thus the movement as a whole benefits from regular communication and networking among its diverse CMOs and from making occasional alliances to issue public statements, hold mass actions and coordinate different strategies by different CMOs.

Diversity of ideologies, strategies and tactics also brings tensions and challenges. These can be reduced to some extent by inviting all members of the movement to subscribe to a set of broad common principles. In the USA, some CMOs have subscribed to the Wingspread Principles on the US Response to Global Warming (see box 5.2). Personally, while accepting the Wingspread Principles, I would like to see CMOs going beyond them and agreeing to the broad policies listed in chapter 4 in the 'Key government policies needed' section. However, it would be counter-productive for the climate action movement to try to insist that all its member-groups have identical policies and tactics. This would destroy the diversity that is one of the movement's strategic advantages.

BOX 5.2

The Wingspread Principles[27]

We, the undersigned organizations and individuals, believe that the United States must take immediate, comprehensive action against global warming, guided by these principles:

- **Urgency**: Global warming is real and it is happening now. Every year that we delay action to reduce emissions makes the problem more painful and more expensive – and makes the unavoidable consequences more severe. Leaders in government, business, labor, religion and the other elements of civil society must rally the American people to action.

- **Effective Action**: The US must set enforceable limits on greenhouse gas (GHG) emissions to significantly reduce them within the next 10 years, and should work with other nations to achieve a global reduction in absolute GHG emissions of 60–80 per cent below 1990 levels by midcentury. Experience proves that voluntary measures alone cannot solve the problem. Aggressive government action, including mandates based on sound science, is imperative and must be implemented now.

- **Consistency and Continuity of Purpose**: Climate stabilization requires sustained action over several decades to achieve deep cuts in greenhouse gas emissions throughout the economy. With its frequent changes of leadership and priorities, however, the American political system does not lend itself to longterm commitments. Leaders in both government and civil society must shape policies and institutions that ensure sustained climate protection.

- **Opportunity**: Mitigating and adapting to global warming offer the opportunity to create a new energy economy that is cleaner, cheaper, healthier and more secure. We must awaken America's entrepreneurial spirit to capture this opportunity.

- **Predictability**: Measures that signal investors, corporate decision makers and consumers of the certainty of future reductions are essential to change the economy.

- **Flexibility**: Deep cuts in greenhouse gas emissions demand and will drive innovation. Our economy will innovate most efficiently if it is given the flexibility to achieve ambitious goals through a variety of means, including market based incentives and/or trading.

- **Everyone Plays**: Measures to stabilize the climate must change the behaviors of business, industry, agriculture, government, workers and consumers. All sectors and the public must be engaged in changing both infrastructure and social norms.

- **Multiple Benefits**: Actions to stabilize, mitigate or adapt to global warming should be considered alongside other environmental, economic and social imperatives that can act synergistically to produce multiple benefits – for example, 'smart growth' practices that conserve forests and farmland while reducing the use of transportation fuels. Many actions to stabilize climate offer local, regional and national, as well as global, benefits.

- **Accurate Market Signals**: The true and full societal costs of greenhouse gas emissions, now often externalized, should be reflected in the price of goods and services to help consumers make more informed choices and to drive business innovation. Policymakers should eliminate perverse incentives that distort market signals and exacerbate global warming.

- **Prudent Preparation**: Mounting climatic changes already are adversely affecting public health and safety as well as America's forests, water resources, and fish and wildlife habitat. As the nation works to prevent the most extreme impacts of global warming, we also must adapt to the changes already underway and prepare for more.

- **International Solutions**: US government and civil society must act now to reduce their own greenhouse gas emissions, regardless of the actions of other nations. Because greenhouse gas emissions and the effects of climate change are global, however, the ultimate solutions also must be global. The US must reengage constructively in the international process.

- **Fairness**: We must strive for solutions that are fair among people, nations and generations.

Organisation of groups

In this section, I'm using the word 'organisation' to describe the act of organising. This contains both the standard meaning – the creation of a structure for a group – and a special meaning used by some US activists such as Saul Alinsky and the Midwest Academy for the activities of an organiser (described below), which go beyond the standard meaning to include planning a campaign. Both degrees of organisation are important for individual CMOs, so that they can build up membership, spirit, trust, resources and capacities of their members and the group as a whole to develop and carry out a campaign plan. CMOs are working for a better society and so their own structure and operations should be designed to achieve this.

The structure to be chosen for the groups depends on four key elements:

- the group's purpose or function;
- whether the group's members are individuals or other groups;
- the group's geographic extent: local, state, national or international; and
- sources of funding: membership subscriptions, philanthropic donors, grants, consulting work, service delivery, or organising conferences/workshops.

Most CMOs receive most of their revenue from membership subscriptions, while some receive occasional large amounts from holding conferences. These are untied sources, which allow flexibility in structure and function. However, groups that perform consulting or deliver services for government or business may have their organisational structures and rules specified and are likely to be constrained from criticising their clients. Furthermore, the tax status of groups that receive government grants may exclude them from lobbying government.

The important thing is to have a structure and rules that are compatible with the above four elements and do not introduce inconsistencies and internal tensions. For instance, problems often arise in groups whose members are both groups and individuals, or in groups that try to combine a service function with activism.

Sociologist, Abigail A Fuller, has reviewed the sociological literature, finding that a generally strong, enduring organisational structure contributes to success.[28] This conclusion is based on a study of the successes and failures of 53 social movement organisations in the USA that operated between 1800 and 1945[29] and another study of homeless movement organisations in 12 cities.[30] Although there is not a large body of recent literature on this topic, Fuller's work confirms my personal experience as a member over several decades of social change groups in areas of environmental protection, appropriate technology and peace. Medium-sized and large groups, that are unstructured with unclear responsibilities and completely informal procedures, rarely make any progress and can become quite stressful and unsatisfying for their members. However, groups that are very small (less than about eight active members) and have limited goals sometimes operate very effectively without formal structure.

For example, when I was active in the peace movement in the early 1980s, I was a member of a small group, Canberra Peacemakers, which specifically promoted social defence, that is, nonviolent community resistance to invasion or internal repression.[31] Normally about six members attended meetings and all decisions were made by consensus. Meetings were harmonious and the group functioned well. At the same time I also belonged to Canberra Program for Peace Committee, a larger organisation, with broad goals, loose organisational structure and chaotic procedures in meetings. From meeting to meeting, people attending changed substantially and the newcomers were not required to respect the decisions of previous meetings or the agenda of the present meeting. Although there were many fine people in the committee, I (and others) found the meetings stressful and progress was very slow.

For groups of 20 people or more, a well-defined structure, with elected office-bearers and formal procedures has the advantages of an assigned set of responsibilities and standard, efficient ways of handling routine tasks. Each meeting can build upon the decisions made at previous meetings. However, excessive formality may have the disadvantages of imposing a constraint on flexibility, creativity and nonviolent direct action. A group that is too tightly organised may disempower the majority of its members. Ideally, a CMO should try to combine the benefits of a formal structure and basic rules of procedure with the flexibility to encourage members to take on significant roles in campaigns and meetings.[32]

My personal preference is for a semi-structured group with, as a bare minimum, a convenor or organiser (discussed below), and elected treasurer, secretary and publicity officer. The elected officer-bearers can mentor volunteers from the group as trainees for these positions. At each meeting there can be several rotating positions:

- a facilitator, who is chosen before the meeting, to allow time for preparation;
- a greeter to welcome new members or volunteers;
- a note-taker, who is not necessarily the secretary;
- a time-keeper; and
- members who have volunteered prior to the meeting to make short presentations on the various ideas, programs, projects, reports and issues for decision.

Thus opportunities arise for a number of people to play active roles in meetings.

The purpose of a CMO is to achieve its goals, not to hold meetings. Therefore, each meeting must have a clear purpose, be well prepared, well facilitated and focused. You do not simply hold a monthly meeting because that is specified in the group's constitution. The principal purpose must be to make decisions that advance the work of the group. These decisions could be on the development of strategy, the planning of tactics or an event, recruiting members or volunteers, evaluating goals and programs, and raising funds.

A good meeting, that gives members a feeling of achievement, needs proper preparation. Before each meeting an agenda is circulated and the issues (pros and cons) are set out for decision on each agenda item. At the meeting, after a circle of introductions, the agenda is discussed and times are allocated to each item. Geography permitting, meetings are preferably face-to-face, otherwise phone or electronic links are needed.

Decision-making is by consensus as first preference. If, after adequate discussion, consensus fails, a majority vote of (say) 75 per cent of members present at a meeting is required to pass a motion. An invitation is given to dissenters to have their positions recorded and to propose at the next meeting how their concerns could be addressed. For small groups, complete consensus is the preferred means of decision-making. The meeting finishes on time.

This approach is most suitable for regular meetings of groups with 8–20 members. Although some CMOs may have several hundred members, it is rare that more than 20 will attend regular meetings. These become a de facto core group, which communicates its decisions to the full membership by regular newsletter, electronic or otherwise. If the core group becomes large and unwieldy, the members can elect a leadership team.

The sociological literature suggests that leadership teams of people from diverse backgrounds and experiences often produce better strategies. In particular, it is valuable to have a mix of people who have been involved in the issue for several years and specialist expert advisers from outside the issue, such as a lawyer or an economist. Although the media and the public tend to see great leaders standing on their own, in fact both Martin Luther King Jr and César Chávez surrounded themselves with diverse teams of advisers.[33]

A really large group may increase its flexibility by breaking up into a number of semi-autonomous groups whose representatives (rotating) meet face-to-face on an annual or biannual basis. The group may also create small, unstructured subgroups of two to four members to work for fixed periods on specific tasks, such as researching an issue or organising an event (for example, lobbying, fundraising, or planning a nonviolent direct action). The subgroups report back at each group meeting. This encourages creativity and flexibility, while maintaining oversight by the main group. When subgroups are composed of a mix of experienced and less experienced members, they can perform a training role in addition to their specific purpose.

Participatory group processes for sharing knowledge and skills are an essential part of building up mutual trust and the capacity of the group to foster social change. In particular, the group can from time to time conduct training sessions on such skills as media advocacy, letter-writing, group facilitation and nonviolent direct action. Social events, such as dinners, dances and celebrations of achievements should not be overlooked. The books *Resource Manual for a Living Revolution*, *Organizing for Social Change* and *Strategy for a Living Revolution* are excellent references for group processes (see 'Key readings and websites').

Role of organiser

Most successful CMOs do not grow and take effective actions without at least one guiding spirit, the organiser. In the context of social change movements, an organiser is a person who facilitates a community to empower itself. (S)he does this by:

- guiding the formation and growth of one or more CMOs;
- helping the group to develop a shared vision, strategy and tactics;
- fostering a democratic group structure and decision-making processes;
- nurturing leaders; and
- guiding the organisation's public meetings, workshops, study groups and actions.

Thus the organiser is somewhat different from a leader, whose role is to lead people into action. The organiser brings individuals together, helping them to build organisations and campaigns. This person facilitates from behind the scenes, while the leader is up front in the public eye.

Organisers of many small community groups work in a voluntary capacity. However the paid organiser has a well-recognised role in the trade union movement and the concept was broadened to the wider social justice movement by the organiser and activist Saul Alinsky,[34] among others. Nowadays the organiser has an increasingly important role in the environmental movement as well.

Motivating people

People have a range of motivations for joining community organisations: altruism, social support, emotional concern, friendship networks and self-interest may all play a role. However, self-interest is less obvious for motivating action on climate change than for action to improve working and living conditions directly. Under these circumstances, how does a CMO jolt people out of their complacency and encourage them to become members, preferably active ones?

One way is to expose the weak policies of the government and the Opposition, demonstrating the huge gap between the outcomes demanded by science and the 'politically feasible' responses of the power-holders. Appealing to people to think about the kind of world they are leaving to their children and grandchildren is an approach that carries weight with parents. Some CMOs are pointing to the loss of icons that many people can identify with, even at a distance: the melting of Arctic ice cap and the glaciers of the Himalayas, and the death of the Great Barrier Reef. Some of the most successful CMOs attract members by making their activities creative, interesting and fun (see appendices 1 and 2).

Many local climate action groups have gained members by simply appealing to people's better nature and by creating an active group of interesting people that others want to join. Friendship networks are often a basis for joining a group. Conversely, people can make new friends and gain social support for their concerns by joining a group. Based on a survey of members of a small climate action group in a suburb of Sydney, Australia, Climate Action Coogee (CAC), researcher Nina Hall identifies both altruism and social support as motivations for joining. On altruism, Hall, Taplin and Goldstein conclude that:

> Given that climate change impacts are not currently being significantly experienced in Coogee nor are likely to impact directly on members' lives as significantly as those living in poorer areas in Australia or in low-lying countries, the values underlying many CAC members' environmentalism most likely stem from altruism … The moral norms that support altruism can also be applied to non-human objects as 'biospheric values'. The pro-environmental behaviour that motivated Coogee citizens to become CAC members was most likely a combination of knowledge and awareness about climate change, altruistic and biospheric values, and emotional concern. Together, these attributes create a complex of 'environmental consciousness' (Kollmuss & Agyeman, 2002, p 256).[35]

Furthermore, on social support, Hall and colleagues conclude that:

> It appears that the social support developed through working together on the Climate Protection Bill project was a significant incentive for CAC coparticipants to become, and stay, involved. This social support developed from the common beliefs and values and a 'sense of belonging' with others who held similar concerns. In supporting each other, we validated our own concerns and our commitment to achieve the aims of the project. One coparticipant commented: '[Working with CAC on the Bill has] reinforced my perception of the empowering nature of a community group, and the joy of working with a highly motivated and skilled group to jointly reach an important goal' … This sense of belonging is often a motivating and sustaining force in groups such as CAC. As Kemmis and McTaggart (2005, p 571) observed, coparticipants provide 'critical support [to one another] for the development of personal political agency and critical mass for a commitment to change'.[36]

As a campaign develops, members of a CMO will need a framework for understanding the various stages they are experiencing. One that has been used effectively by activists to describe several quite different campaigns is the Movement Action Plan.

Movement Action Plan

The Movement Action Plan (MAP) provides activists with a broad framework and strategic analysis of the way nonviolent social movements can achieve social change.[37] Bill Moyer and co-authors outline the plan:

> It shows how successful social movements typically travel along similar long and complex roads, which usually take years or decades. MAP allows activists to –
>
> - identify where, on the normal eight-stage road of movement success, their movement is at any specific time;
> - create stage-appropriate strategies, tactics and programs that enable them to advance their movement along to the next step on the road to success;
> - identify and celebrate their movement's incremental progress and successes;
> - play all four roles of activism effectively;
> - overcome irrational feelings of powerlessness and failure; and
> - engage ordinary citizens in the grand strategy of effective social movements – participatory democracy.[38]

MAP was first published in 1987 by Bill Moyer, a social change activist and founding member of the Movement for a New Society. The most recent version of MAP, with examples from several issues, was published in 2001 in a book by Bill Moyer and several co-authors. MAP classifies a campaign by a social movement into the eight stages (shown in table 5.2).

TABLE 5.2 The eight stages of MAP

Stage	Brief description
1. Normal times	• A critical social problem exists that violates widely held values. • Power-holders support problem. Their official policies express widely held values, but the real operating policies violate those values. • Public is unaware of the problem and so supports power-holders. • Problem/policies not a public issue.
2. Prove the failure of official institutions	• Form many new local opposition groups. • Use official channels – courts, government channels commissions, hearings, etc – to find and demonstrate that they don't work. • Become experts by doing research.
3. Ripening conditions	• Recognise problem and number of victims grow. • Public sees victims' faces. • More active local groups form. • Protests take place. • 20–30% of public opposes power-holder policies.
4. Take-off	• Trigger event. • Dramatic nonviolent actions/campaigns take place. • Actions show public that conditions and policies violate widely held values. • Nonviolent actions are repeated around country. • Problem put on social agenda. • New social movement takes off rapidly. • 40% of public oppose current conditions/policies.
5. Perception of failure	• See goals unachieved. • See power-holders unchanged. • See numbers down at demonstrations. • Despair, hopelessness, burnout, droput, seems movement ended. • Emergence of negative rebel.
6. Majority public opinion	• Majority oppose present conditions and power-holder policies. • Show how problem and policies affect all sectors of society. • Involve mainstream citizens and institutions in addressing problem. • Problem put on political agenda. • Promote alternatives. • Power-holders demonise alternatives and activism. • Counter each new power-holder strategy. • Promote paradigm shift, not just reforms. • Retrigger event happens, re-enacting stage 4 for a short period.
7. Success	• Large majority oppose current policies and no longer fear alternatives. • Some power-holders still try to make minimal reforms, while movement demands social change. • Many power-holders change position. • End-game process: power-holders change policies or lose office. • New laws and policies.
8. Continuing the struggle	• Extend successes (eg, even stronger civil rights laws). • Oppose attempts at backlash. • Promote paradigm shift. • Focus on other subissues. • Recognise and celebrate successes so far.

SOURCE Adapted from Moyer et al (2001), figure 1, with permission.

The other main feature of MAP is that it recognises four roles for social movements and the social activists who comprise the movements:

1   *Responsible citizen:* Supporting fundamental values of a good society.
2   *Rebel:* Opposing social conditions and policies that violate those values.
3   *Change agent:* Educating, organising and involving citizens to oppose present policies and seek constructive solutions.
4   *Reformer:* Working with official, judicial and political institutions to transform solutions into new laws and policies, and to gain social acceptance of them.

As the campaign proceeds through its eight stages, the emphasis upon the different roles shifts. For instance, the 'rebel' and the 'change agent' are essential in stages 2–6. The 'reformer' has a role in stage 2, to show the failure of official institutions, and in stages 7 and 8, to formulate and implement the new policies, laws and organisational structures. The 'responsible citizen' is needed in all stages.

Different CMOs may also concentrate on different roles. For instance, it is difficult for a small climate action group to be simultaneously lobbying the government while attacking its policies in the media. However, a large environmental NGO can and in some cases should have different members playing both roles. The 'good guy, bad guy' tactic can be successful at times.

MAP has been used to analyse the social movements against nuclear energy, for US civil rights, for US gay and lesbian rights, for breast cancer action and against globalisation.[39] In applying MAP, this chapter divides the climate change issue into two linked subissues:

1   gaining acceptance of greenhouse science and hence the seriousness and urgency of the problem; and
2   developing and implementing an effective national greenhouse response strategy.

MAP stage for acceptance of greenhouse science

As discussed above, several public opinion polls show that the vast majority of the Australian public recognises the seriousness and urgency of global climate change.[40] However, neither the Labor Government nor the Coalition Opposition has been treating the issue as urgent (see chapter 2). Therefore, we suggest that the Australian campaign has been in stage 6 'Majority public opinion' during the latter part of 2007 and throughout 2008. It may require a dangerous trigger event, such as the partial collapse of the West Antarctic Ice Sheet *and* a significant abrupt rise in sea-level, to engage the power-holders and move into stage 7 'Success'.

In the USA, it is unclear whether majority public opinion recognises that anthropogenic global climate change is a serious and urgent issue. According to one poll, three-quarters of Americans were either 'much more convinced' or 'somewhat more convinced' that global warming is happening, compared with their beliefs in 2004.[41] In contrast, a Pew Center poll conducted a few months earlier found that, while 74 per cent say that climate change is a serious or very serious problem, only 41 per cent of Americans believe that there is solid evidence that human activity is causing it. Unlike Australians, Americans considered the issue to be a low priority.[42] This doubt has been attributed to the success of the anti-climate action groups (see below), especially the conservative think-tanks. It appears that the campaign to gain acceptance of greenhouse science in the USA in 2008 had reached stage 3 'Ripening conditions'. The election of President Obama could possibly lift this to stage 4 'Take-off' in 2009.

MAP stage for climate action

In Australia, the campaign to gain acceptance of a coherent and effective set of mitigation policies and strategies is at an earlier stage than the campaign on greenhouse science. Although the problem is widely recognised and numerous climate action groups exist and agree on the basic strategies, the general public is confused about the choice of responses. Some have been diverted from effective strategies by the intense marketing campaign for nuclear energy. My previous book *Greenhouse Solutions with Sustainable Energy* shows that nuclear energy, based on existing commercial technologies, cannot contribute significantly to the solution to global warming. Others believe incorrectly that they can solve the problem simply by reducing their own direct greenhouse gas emissions (for example by changing light-bulbs, purchasing Green Power and driving less).[43] Others accept the claims of neoclassical economists that the forthcoming emissions trading schemes will be sufficient for reducing emissions, provided the target is adequate.[44] Others accept the notion, propagated by climate change deniers, that a reduction in Australia's emissions would not influence other countries such as the USA, China and India to cut their emissions. Therefore, I suggest that the campaign for climate change mitigation, in Australia in early 2009, is well into in stage 3 'Ripening conditions'. This is a little less optimistic than stage 4 'Take-off' result obtained by Nina Hall and Ros Taplin for the whole of the climate action movement by using a different method.[45]

In the USA, a large minority of state governments and about 400 municipal governments have set greenhouse mitigation targets. Under the Regional Greenhouse Gas Initiative, ten eastern and mid-Atlantic US states have commenced an emissions trading scheme that will cap and then reduce CO_2 emissions from the power sector by 10 per cent by 2018.[46] President Obama is committed to a national cap-and-trade emissions trading scheme. He has promised US$150 billion over ten years to renewable energy, but justifies this more in terms of energy security than global warming. If we take these actions as an indicator of public opinion, then the campaign to implement effective greenhouse solutions in the USA may be reaching stage 3 'Ripening conditions'.

MAP describes in broad terms the development of a national or international campaign by the climate action movement. Next we examine the nature of this movement.

Climate action movement

Types of climate movement organisation

Table 5.3 classifies CMOs into various types and lists some examples.

Among the *International, National and State Generalist NGOs* with climate campaigns, there are environmental organisations – such as the Friends of the Earth, Greenpeace and WWF – and social justice groups such as Oxfam International and, in Australia, the Brotherhood of St Laurence. In these organisations, climate action is just one of several campaign areas. As expected, the social justice groups have particular concerns about the equity aspects of climate change and response strategies (see chapter 1).

Climate action groups are groups whose principal goal is to reduce anthropogenic climate change substantially. This category comprises a few large climate action groups that are generally funded by philanthropic foundations and many small groups that are funded by membership subscriptions. For example, the Pew Center on Global Climate Change is a large US-based NGO that

> brings together business leaders, policy makers, scientists, and other experts to bring a new approach to a complex and often controversial issue. Our approach is based on sound science, straight talk, and a belief that we can work together to protect the climate while sustaining economic growth.[47]

Over 100 climate action groups in Australia are listed on the network website <http://www.climatemovement.org.au>. Another 60 or so are members of Climate Action Network Australia <http://www.cana.org. au> (discussed below). The vast majority of climate action groups have members who are geographically local and campaigns that are focused predominantly in their respective local regions. A few, such as the Pew Center and the Climate Institute[48] operate on a national scale.

One of the few *business alliances* devoted to climate action is the Australian Business Roundtable on Climate Change, discussed in box 5.1. An important coalition of trade unions is the Australian Council

TABLE 5.3

Classification of climate movement organisations

Type of organisation
International, national and state generalist NGOs with climate campaigns
Climate action groups
Business alliances and business-NGO alliances
Networks of CMOs
Trade unions with concern about climate action
Professional organisations
Faith groups
Students
Local government
Academic/research
'Left' discussion groups
Individuals

SOURCES From the author's knowledge of the literature, including websites, and personal contacts.

NOTES This list is not intended to be comprehensive. However, it shows the diversity of types of group and of groups within a category. Some CMOs are deliberately listed in more than one category.

Examples of CMOs and countries	Website
Australian Conservation Foundation (Aus)	http://www.acfonline.org.au
Avaaz.org (international)	http://www.avaaz.org
Brotherhood of St Laurence (Aus)	http://www.bsl.org.au
Friends of the Earth (many countries)	eg, http://www.foe.org
GetUp (Aus)	http://www.getup.org.au
Greenpeace (many)	http://www.greenpeace.org/ international
Oxfam International (many)	http://www.oxfam.org
Sierra Club (USA)	http://www.sierraclub.org
WWF (many)	http://www.wwf.org
Clean Energy for Eternity (Aus)	http://austcom.org.au/cefe.html
Climate Institute (Aus)	http://www.climateinstitute.org.au
Hepburn Renewable Energy Association (Aus)	http://www.hrea.org.au
Kingsnorth Climate Action Medway (UK)	http://nonewcoal.org.uk/?q=node/39
Mt Alexander Sustainability Group (Aus)	http://masg.org.au
Pew Center on Global Climate Change (USA)	http://www.pewclimate.org
Rising Tide (Aus)	http://www.risingtide.org.au
Australian Business Roundtable on Climate Change (Aus)	http://www.businessroundtable.com.au
CERES (USA)	http://www.ceres.org
Apollo Alliance (USA)	http://apolloalliance.org
Australian Student Environment Network (Aus)	http://asen.org.au
Clean Energy Council (Aus)	http://cleanenergycouncil.org.au
Climate Action Network (many)	http://www.climatenetwork.org
Stop Climate Chaos Coalition (UK)	http://www.stopclimatechaos.org
www.climatemovement.org.au (Aus)	http://www.climatemovement.org.au
Australian Council of Trade Unions (Aus)	http://www.actu.asn.au
Australian & New Zealand Solar Energy Society (Aus)	http://www.anzses.org
Environment Institute of Australia & New Zealand (Aus & NZ)	http://www.eianz.org
Faith and Ecology Network (Aus)	http://www.columban.org.au/Our-works/peace-ecology-justice/faith-ecology-network.html
Australian Religious Response to Climate Change (ARRCC) (Aus)	http://www.arrcc.org.au
Australian Student Environment Network (Aus)	As above
Central Victorian Greenhouse Alliance (Aus)	http://www.cvga.org.au/main
Centre for Energy & Environmental Markets, UNSW (Aus)	http://www.ceem.unsw.edu.au
Institute of Environmental Studies, UNSW (Aus)	http://www.ies.unsw.edu.au
Politics in the Pub (Aus)	http://www.politicsinthepub.org.au
Search Foundation (Aus)	http://www.search.org.au
James E Hansen	http://www.columbia.edu/~jeh1

of Trade Unions (ACTU).[49] It has commissioned a report with the Australian Conservation Foundation that finds a huge potential for 'green collar' jobs in several sustainable development areas including energy efficiency and renewable energy.[50] The ACTU is also one of the members of the Southern Cross Climate Coalition (see box 5.1), which has called for the expansion of the Mandatory Renewable Energy Target.

Coalitions, alliances and networks of CMOs

The Climate Action Network (CAN) is a worldwide group of over 430 NGOs working to promote government and individual action to limit human-induced climate change to ecologically sustainable levels. CAN members work to achieve this goal through the coordination of information exchange and NGO strategy on international, regional and national climate issues. There are regional offices in all the inhabited continents.[51]

In 2005, Stop Climate Chaos, a coalition of environmental, development, trade union, faith and women's groups was formed in the UK. According to its website, the coalition has over 70 member organisations with a total supporter base of four million people.[52] It coordinates national actions and provides climate change news from the UK and the European Union. Another large network of UK climate action groups is Campaign Against Climate Change. It organises the annual National Climate March, coordinates the Global Day of Action and organises marches and demonstrations on specific climate issues, such as opposing the proposed new coal-fired power station at Kingsnorth, the third runway at Heathrow and unsustainable biofuels.[53] Appendix 1 outlines the tactics and actions of some UK CMOs that belong to these networks and appendix 2 outlines the tactics and actions of some US CMOs.

Australia has several climate action alliances and networks including the following.

Climate Action Network Australia (CANA) is the Australian branch of the global Climate Action Network mentioned above.

Despite its name, CANA is actually better described as an alliance than a network. It comprises over 60 local, state and national environmental, health, community development, and research groups from throughout Australia. As well as networking, participating in international climate negotiations and providing news and information to its member groups, CANA campaigns on behalf of its member groups on several climate change issues, such as opposition to new conventional coal-fired power stations and speeding up the implementation of the expanded Renewable Energy Target.[54]

ClimateMovement.org.au is 'a hub for Climate Action Groups around Australia to connect with each other, access resources, share success stories and collaborate to make a more powerful and effective climate movement'.[55] It has over 100 climate action groups on its website. It is coordinated by a medium-sized, state-based CMO, the Nature Conservation Council of New South Wales.

Other Australian networks are the Australian Youth Climate Coalition, whose activities include lobbying politicians, producing a cinema advertisement, preparing educational materials, training its members and public speaking with a focus on youth as the recipients of a damaged climate future,[56] and the Australian Student Environment Network, which is made up of campus environment collectives from all over the country.[57]

Anti-CMOs

A small number of groups are opposed to strong action to reduce greenhouse gas emissions and could therefore be classified as 'anti-CMOs'. Some of these organisations are open about their goals, while others claim to be impartial or even pro-environment. The formation of front organisations for powerful vested interests is known as '*astro-turfing*' and has been discussed by Sharon Beder, visiting professor in the School of Social Sciences, Media and Communication at the University of Wollongong.[58]

The former US-based organisation, Global Climate Coalition, was openly a peak organisation for deniers. It is discussed in chapter 2.

Another vocal, open opponent of strong action to reduce emissions is the Australian Workers Union, whose members are drawn from a wide range of some of the biggest greenhouse gas emitting industries: pastoral and agricultural, aluminium, aviation oil and gas, mining, construction and steel.[59] Because most other trade unions don't directly contradict it publicly, presumably to maintain union solidarity, the media sometimes describe it incorrectly as the voice of the Australian trade union movement on this issue. In reality, members of the trade union movement hold a wide range of views on climate change and climate action.

The ACTU is mentioned above. The Construction, Forestry, Mining and Energy Union supports action to reduce greenhouse gas emissions, by an expansion of renewable energy and carbon capture and sequestration (an unproven technological system).[60] In 2003, John Maitland, then national secretary of the CFMEU, stated:

> So there is no point in trying to scare this union about the employ-
> ment impacts of Kyoto targets ... a significant job creator it [coal
> mining] isn't. Very few mineworkers expect or hope for their kids to
> work in coal-mines. The job opportunities lie elsewhere.[61]

CFMEU supports both the expansion of the Mandatory Renewable Energy Target and the construction of a new conventional coal-fired power station in New South Wales by 2013.[62] So I'm unsure whether to classify it as an anti-CMO with renewable energy sympathies, or as simultaneously a CMO and an anti-CMO.

Conclusion

For a movement seeking to achieve nonviolent social change, strategy and organisation are vital. Strategy is the planning and conduct of long-term campaigns to achieve broad goals. Effective strategy brings the power of the movement to bear on institutions, organisations and individual power-holders. Strategic planning is essential for both individual climate movement organisations and for coalitions, alliances and networks of CMOs. It is at least as important for nonviolent social change as for a military campaign.

Organisation enables the number of members of the movement to be built up and then mobilised and concentrated upon the targets of the campaigns. This is the great potential strength of the climate action movement. Although it cannot match the economic power of the vested interests in greenhouse gas emissions, it can mobilise far greater numbers of people, has sufficient diversity to influence almost all parts of society, and its members will generally have much stronger convictions in their cause. In a democratic society and even in some dictatorships, large numbers of people with conviction can ultimately win, but only if they are organised and follow a well thought-out strategy.

For a small, straightforward campaign or issue that is part of the larger climate action campaign, SWOT analysis may be adequate as a strategic planning tool. For planning a larger, more complex campaign, I find the Midwest Academy's Strategy Chart valuable. For tracking and understanding the progress over time of the whole campaign on a large scale, the Movement Action Plan is a useful resource.

The climate action movement has diverse organisations and groups, each having their own strengths. Provided they cooperate from time to time on state, national and international actions, they have the collective potential to influence all sectors of society. The tactics available to these diverse groups are discussed in the next chapter.

6

WINNING CAMPAIGN TACTICS

Franklin D Roosevelt once said, 'Okay, you've convinced me. Now go out there and bring pressure on me.' To achieve social change, clearly we have to mobilise the support of the majority of the population. But we also need to ensure that the power-holders understand the issues and necessary policies. If we lobby without grass-roots support, this will carry little weight with power-holders. On the other hand a mass movement will not gain the right decisions from the power-holders unless it has communicated its requests to them, usually by lobbying. For this reason, the climate action movement needs a wide range of tactics from different climate movement organisations. Some tactics can be directed towards influencing the power-holders directly, others towards the community at large and various key subgroups within it, and still other tactics towards important intermediaries such as the media.

Most actions will be 'soft', in the non-confrontational sense. They are designed to win people around to the value of and need for the proposed changes. However, not all power-holders can be won over by soft words, especially if they identify strongly with the vested interests that maintain the status quo. There are still some politicians and business leaders in key positions who are actively opposing climate action. So, some actions will have to be confrontational, such as debate in public meetings and the media, demonstrations, pickets and boycotts. These nonviolent confrontations are not intended to win over diehard power-holders, but rather to resist the latter's bad decisions, educate the public and create media events.

This chapter outlines the principal tactics available to climate movement organisations (CMOs).

To carry out effectively most of the tactics outlined here, a CMO should take responsibility for training its members. This can be done either by a good organiser, or by bringing in an experienced external agency or facilitator. Some examples are the Change Agency and Unfolding Futures in Australia; and the Midwest Academy, the Ruckus Society and the Rockwood Leadership Program in the USA.[1]

CMOs should keep in mind that just because an action is

nonviolent, that doesn't necessarily mean it is lawful. The consequences of conducting unlawful actions can range from being caught up in expensive litigation, to being fined, or even imprisoned. When considering the tactics discussed in this chapter, you should first consider the legal implications of any proposed action, and seek legal advice. Handbooks such as *Campaigning and the Law in New South Wales*, published by the Environmental Defender's Office,[2] can help you identify potential legal issues. However, they are not a substitute for legal advice.

Before discussing individual tactics, the CMO must consider how to *frame* the issue or campaign, that is, how to present it to the public in a context and language that you choose.

'Framing' the campaign

Communication is a vital part of any campaign. You must decide how you wish your group and your campaign to be seen by others, including decision-makers, influential organisations such as business, trade unions and professionals, other NGOs, the media and the community at large. To some extent you can influence your 'image' by the way you describe your group and its vision, its goals, its campaign and the issues as you see them. Your website, press releases, media interviews, published articles and communications with individuals all help to do this. This process is known as 'framing' and it involves choices of conceptual framework for the campaign and the language used.

Opponents, the media and other interested parties each have their own agendas and each will try to place you and your issues into a 'box' of their own making. In other words, they will try to 'frame' you. For example, it is common to label individuals and groups who are pushing for constructive social change as 'emotional', 'ideologically driven', 'irresponsible' and even 'unscientific'. Of course, much of this is projection in the psychological sense. There is no-one more emotional than the CEO of a polluting industry who is facing the higher costs of environmental protection or reduced profits. There is no-one more ideologically driven than a narrow neoliberal economist who tries to impose the competitive market model on parts of the economy that are manifestly suffering from market failure. No-one is more unscientific than a climate change denier who claims, contrary to all the empirical evidence, that the majority of glaciers are actually expanding instead of shrinking.[3] Nevertheless, you must be ready to 'reframe' your group, yourself and the campaign at every opportunity.

Environmental campaigner and communications consultant Chris Rose devotes a whole chapter of his book to 'constructing campaign propositions'. He defines a 'campaign proposition' as a summary of 'what the campaign is about'. He suggests that an effective campaign proposition is one that states simply the problem, the solution and the benefit.[4] As an example, consider the following hypothetical media interview:

Interviewer: Why is your group campaigning for gross feed-in tariffs for solar power?

Campaigner: The problem is that there is little economic incentive in this country to build up the market for solar power, which has enormous potential but is still expensive.

Interviewer: So what's the solution?

Campaigner: Gross feed-in tariffs pay a premium price for all electricity that is generated on a rooftop by solar energy and sold back to the electricity grid. The tariffs are financed by a tiny increase in the overall price of electricity.

Interviewer: Who benefits from gross feed-in tariffs and why does your press release reject *net feed-in tariffs?*

Campaigner: Gross feed-in tariffs build up the market for solar power and bring down its price. They benefit both the solar industry and everyone who wants to reduce greenhouse gas emissions. We reject net tariffs because they only pay the higher tariff on the difference between the solar power generated and the grid electricity used. For many households, there is no difference and so there is no incentive to purchase solar electricity. That's why we are backing *gross feed-in tariffs*.

Rose suggests that the campaign proposition can be made even stronger in some cases by adding a component on who is responsible for the problem and what actions are needed in addition to the solution to the immediate problem. For example, continuing the previous interview:

Interviewer:	How do you justify solar electricity receiving the subsidy of a feed-in tariff?
Campaigner:	For decades the production and use of fossil fuels have received government subsidies of over A\$10 billion per year in Australia,[5] while renewable energy has been starved. Gross feed-in tariffs offer electricity users a means of building up the solar market by paying a little extra on their electricity bills. It won't cost the government anything!
Interviewer:	What other actions do you think should be taken to set things right?
Campaigner:	Fossil fuels are a mature industry and do not deserve continuing subsidies. It's unjust and a waste of taxpayers' money. We would like listeners to ask their MPs to lobby the government to phase out subsidies to these wealthy industries. By the way, in Germany the magnitudes of the feed-in tariffs to renewable energy are being reduced every year, so they are not seen as permanent.

Note that in her last response the campaigner is following the advice from the Movement Action Plan stage 2 to show that current practice of subsidising the rich fossil fuel industries violates widely held values.

To demonstrate the framing of a campaign proposition, this interview was conducted in neutral language. However, in the real world, interviewers, interviewees, proponents and opponents all choose words, phrases, similes, metaphors and arguments to create a certain impression. Linguist George Lakoff is an expert in the way language shapes how we think. He points out that

> environmentalists have adopted a set of frames that doesn't reflect the vital importance of the environment to everything on Earth. The term 'the environment' suggests that this is an area of life separate from other areas of life like the economy and jobs, or health, or foreign policy. By not linking it to everyday issues, it sounds like a separate category, and a luxury in difficult times ... [As an alternative] when environmental issues are cast in terms of health and security, which people already accept as vital and necessary, then the environment becomes important.[6]

Climate change campaigners should reframe the neoclassical economics terms 'command and control', using instead the more neutral 'regulations and standards'; and 'burden sharing', using instead 'the distribution of costs' in the context of an emissions trading scheme. 'Clean coal' is a marketing term for 'CO_2 capture and sequestration', a complex technological system that doesn't exist for coal power. Even the term 'global warming' has cosy overtones to someone who is unaware of the current impacts and the disasters that are anticipated. In view of the melting of ice-caps and glaciers, floods, droughts, storms, fires, etc, 'climate disruption' or 'climate crisis' or even 'climate emergency' are more appropriate terms.

Now we are ready to consider the tactics available to CMOs.

Lobbying decision-makers

In this chapter, lobbying is defined to be communication with decision-makers who are addressed specifically by name or position. In this sense, lobbying covers face-to-face meetings, phone-calls, emails, letters, petitions, submissions to government enquiries, addresses to public hearings and accountability meetings. Lobbying is a legitimate and important part of participatory democracy. Since vested interests place considerable resources into lobbying,[7] CMOs need to participate as well.

Decision-makers to be lobbied include:

- ministers, shadow ministers and their advisers;
- other members of parliament in all spheres of government and their advisers;
- senior departmental officials;
- chairs and members of parliamentary committees;
- party committees; and
- CEOs and other senior managers of businesses, peak organisations and other bodies.

It is worth singling out decision-makers who have a particular interest in climate change or who may be sympathetic and willing to push the cause of genuine climate action. It's also worth talking with those who do not have a strong position. Don't waste precious time on those known to be strongly opposed to climate action. The larger CMOs tend to focus their lobbying on federal and state governments, while the smaller CMOs tend to interact with local government and local members of parliament of federal and state governments.

CMOs that are not powerful or influential in their own right may have a hard time arranging to meet a minister or the CEO of a large business. The chances of gaining an interview with such a powerful person will be greatly increased if the lobbyists are representing a large group, such as a national or international environmental, social justice, professional, trade union or business organisation, or represent a joint delegation from several organisations.

Lobbying and other forms of negotiation with power-holders must inevitably involve the exercise of power. Otherwise there is no need for a power-holder even to meet with the group, let alone take genuine action to address the group's concerns. When lobbying a politician, you must demonstrate either your group's influence in the electorate – that your group can increase or decrease votes for that politician or her party – or your group's ability to gain positive or negative media publicity for the politician and her party. When lobbying a business leader, if you can demonstrate your group's capacity to influence his business's sales through consumer action or media publicity, this would strengthen your position. The climate action movement needs the grass-roots support of a large number of people to demonstrate electoral and consumer power and to interest the media. The more supporters that are organised and shown to be active, the more influence the movement has. Otherwise lobbying falls upon deaf ears.

If you have arranged to meet with a decision-maker, consider the purpose of your interview carefully in advance of the meeting and prepare a one-page briefing note to send to the decision-maker before the meeting. Since most meetings with politicians, public officials and CEOs are limited to 30 minutes, decide in advance on clear goals and preferred outcomes of the meeting. It is rarely fruitful to bombard a power-holder with a large amount of information and a vast array of demands. Instead, consider what specifically you want the decision-maker to do on your behalf and how you can follow-up to ensure that this person actually performs an agreed task.

If you are lobbying your local member of parliament (MP) on global warming, ask her to arrange a meeting with an appropriate minister, such as the Minister for Climate Change or the Minister for Energy. If that is not possible, you could ask your local member to forward a letter to the minister, or you could request support from the MP for a local public meeting or a local project to set up a community renewable energy project. Be sure to follow up your meeting with a communication to express thanks and summarise the outcome and what the decision-maker has undertaken to do. It is helpful during the meeting to establish a means of ongoing dialogue on the subject.

Timing is important. Don't arrange a meeting with a politician to take place a few weeks before an election or a budget, but instead get in early before the key decisions have been made. If you are organising a delegation to Parliament House, ensure that parliament will be in session at that time and make appointments to see as many relevant

MPs as possible, allowing sufficient time between appointments for evaluating progress and walking between offices.

All lobbying should be polite and calm. Power-holders are human beings too, and they will switch off if the meeting becomes unpleasant for them. Don't spend a lot of time on the problems, but rather move rapidly to focus on the constructive solutions you desire. Emphasise the benefits to the power-holder of following the course of action you recommend.

Another form of lobbying is letter-writing. This is done most pleasantly and effectively in groups. Members agree on the main points and then write letters in their own respective styles, exchanging drafts to check content, clarity and grammar. Try to keep the length under one page, two at most. Don't send form letters. To obtain the maximum outcome from your efforts, send concise versions of your letters to newspapers and magazines. Letters to the editor should be preferably less than 200 words and make just one or two main points. If possible, link them to a current newsworthy event.

Petitions are a type of mass lobbying. If they can gain a large enough number of signatures, or if they are sent by a group of distinguished people (for example, scientists, academics, leading businesspeople, church leaders, mayors and CEOs of large NGOs), they are a potential source of media publicity. They are more valuable when they contain specific demands rather than generalities, for example:

- a national greenhouse gas reduction target of 30 per cent below the 1990 level by 2020;
- a renewable electricity target of 25 per cent of all electricity by 2020;
- a national gross feed-in tariff for solar power to be implemented within six months.

Be sure to check that your petition has the correct formal wording to be presented in parliament.

Participation in government enquiries and public hearings is both a lobbying process and a way of putting your views on the public record. However, many such events are arranged by power-holders to create the impression of public consultation while they actually have a predetermined conclusion. An obvious example was the Uranium Mining, Processing and Nuclear Energy Review, held in Australia in 2006, in which every member of the review committee was either from the nuclear industry or nuclear research.[8] Therefore, it is rarely worthwhile for a CMO to spend weeks preparing a submission, unless it has multiple uses. It is usually better to make a short submission and offer to expand on it in person if invited.

An accountability session is a large public meeting with decision-makers organised by the CMO. It serves the dual purpose of publicly lobbying the decision-makers on behalf of the participants and further educating the participants about the issue and the political aspects. Naturally, few decision-makers will attend such meetings unless the political costs of their absence are large. The best events for attracting politicians are public meetings of candidates before an election and large conferences where the politician is an invited speaker. In the latter case many politicians try to avoid answering questions.

For a detailed guidebook on lobbying by community organisations with case studies, see RMIT's Advocacy Toolkit.[9]

The tactics and actions of CMOs in the UK and USA are outlined in appendices 1 and 2.

Educational activities

Educational activities include public meetings and teach-ins; conferences, seminars and workshops; media coverage of events, actual or staged; opinion pieces and letters to the editor in newspapers and magazines; door-knocking; reports, leaflets, educational materials for schools, websites, blogs and email newsletters.

All types of CMO conduct educational activities, for their own members, decision-makers and the wider community. An active climate action group may engage locally with government, businesses, media, and clubs and societies (even sporting clubs).[10] An active example is the Australian climate action group, Clean Energy for Eternity, described in box 6.1.

Some of the most successful environmental action groups use a mix of education, nonviolent actions (such as marches and demonstrations), lobbying and networking with other action groups. An example is the Danish anti-nuclear power group OOA, which was originally called The Organization for Information on Nuclear Power and now is The Movement on Energy and the Environment. Formed in 1974, OOA mounted a grass-roots campaign that spanned the whole of Denmark. Its educational activities included leafleting every Danish home in the Winter of 1979–80, followed up by many home visits, and widespread distribution of posters, stickers and badges with the now world famous smiling Sun image with the firm but friendly text 'Nuclear Power? – No Thanks'. OOA was a key player in stopping the strong push to introduce nuclear power into Denmark. Nowadays OOA is working for the closure of Swedish nuclear power stations, one of which is sited only 20 kilometres from Copenhagen, and obtaining and disseminating information about the victims of the nuclear disaster at Chernobyl.[11]

Paid advertising is an expensive means of getting out a message. A cheap and simple means of spreading short messages is to erect signs at highly populated locations such as major road intersections, train and bus stations, fairs and sporting events. Although authorities will remove the signs, that could take several hours or even several days at some highly visible sites that are out of reach of short ladders. This kind of action may not be lawful in some jurisdictions, and penalties such as fines could apply. Seek professional legal advice before trying this.

A key part of any education campaign is for the climate movement to create a small unit that refutes promptly and authoritatively in the media the statements characterised by 'spin' and obfuscation that are disseminated as a matter of routine by governments and vested interests. The kind of statement that deserves exposure, even ridicule, is one made by the Australian Prime Minister Kevin Rudd upon the release of his government's White Paper on emissions trading. Mr Rudd claimed that the scheme is 'balanced' between the requirements of industry and the environment. Environmental organisations responded that Mr Rudd was actually trading off the science of climate change against the demands of vested interests. In the words of economist Regina Betz, joint director of the Centre for Energy and Environmental Markets at UNSW, 'The proposed 2020 targets of emission reductions of 5 to 15% are, according to the climate science, entirely inadequate for an equitable global response to avoid dangerous global warming.'[12] Scientists, economists and environmentalists pointed out that the scheme may be ineffective in reducing emissions, because it sets a tiny emissions target, gives huge 'compensation' to coal-fired electricity generators and emissions-intensive trade-exposed industries, and rebates the CO_2 price on petrol and diesel (see chapter 4). This 'balanced' scheme will give negligible incentive to change the electricity generation mix or to shift transportation away from oil (at least for the foreseeable future).

BOX 6.1

Clean Energy for Eternity

Philippa Rowland, publicity officer, CEFE

Clean Energy for Eternity (CEFE)[13] is a not-for-profit community group based in south-east New South Wales, Australia. It acts as a positive catalyst to inspire practical local initiatives to address climate change. CEFE sees its key role as creating pathways to action and developing high profile models that encourage communities to seize the opportunity to create a sustainable future. There are now active CEFE groups in six shires of the region and in the Sydney suburbs of Mosman and Manly, all committed to CEFE's target of 50 per cent reduction of energy use and 50 per cent renewable energy by 2020. Potential impacts of climate change are very real for the '50/50 by 2020' rural shires, where major economic keystones are tourism and agriculture. The existence of our Sydney CEFE groups opens up important opportunities for urban-rural partnerships in finding workable solutions for climate change.

Clean Energy for Eternity began in May 2006 when local orthopaedic surgeon Dr Matthew Nott invited the community to show its concern for climate change by joining him in making a human sign. Three thousand people, comprising 10 per cent of the local population, turned out on a beautiful autumn day and wrote the words 'Clean Energy for Eternity' along Tathra Beach – and a climate action group was born. CEFE has remained strongly committed to grass roots action as a climate group, as evident in the long list of community activities that have taken place over the past two and a half years.

Human Signs have remained a hallmark of CEFE engagement with the community. Over 50 signs have been created to date, including the combined schools human sign in which students from over 30 schools turned out on school ovals to write their concerns and hopes about climate change.

Keeping actions creative and fun has assisted in maintaining the pace and the enthusiasm. From the involvement of a local landscape sculptor and artist in creating an enormous Eternity Cyclone sculpture out of discarded washing machines, symbolizing the energy we waste in our throw-away society, through to successful art auctions, fashion parades and mural competitions, CEFE has managed to involve the artistic members of the community in a meaningful way.

CEFE has successfully engaged with existing community groups, such as surf life-saving clubs, churches, rural fire brigades and numerous other interest groups. We have a powerful three-way partnership with the Australian Community Network, AustCom, and the Sapphire Producers Association, which has resulted in a high quality, quarterly journal called Sustain, delivered free of charge up and down the coast and inland to Cooma, Thredbo and Canberra.

Thus CEFE has been influential in its advocacy for climate change action. Reflecting on our modest successes, three contributing factors stand out:

- our timing, riding the wave of community concern and converting it into practical positive action;

- our location in the marginal electorate of Eden-Monaro in the lead-up to a federal election;

- our engagement right across the community, bringing credibility as a mainstream movement.

Use of the media has raised our profile and increased our group's political influence. Twice nominated as finalists for the IAG Eureka Award for Innovative Solutions to Climate Change, CEFE has taken every opportunity to act as a catalyst for urgent action on climate change at the individual, community and political level. Formal submissions to government on the Garnaut Report and the Federal Green paper on the Carbon Pollution Reduction Scheme, calling for an Integrated Renewable Energy Policy, have generated letters of acknowledgement from the relevant Ministers but not, as yet, tangible action.

We have been greatly supported by the new Federal MP for Eden-Monaro, the Hon. Mike Kelly, who stated a commitment in his maiden speech to create a '50/50 by 2020' electorate in Eden-Monaro. He was also instrumental in an election commitment to provide funds to assess the feasibility study of a community solar power station for the region.

The group secured $100 000 for a feasibility study for the solar power station under the Federal Government's Green Precincts Program, with a further $1 million if viable. The project will provide a highly visible example of the potential for the region to shift to renewable energy. It will also develop a macro solar investment model that could substitute for, or complement, the existing Australian Government Photovoltaic Rebate Program and accelerate the roll-out of renewable energy generation capacity in Australia. A solar power station provides opportunities for community members to take the initiative and have a direct stake in the development of renewable energy in the region. It also provides a strong focus for our rural-urban partnerships.

Media and other communications

Several benefits can be gained from appropriate media coverage. It is an educational activity for the general public. It can also give your CMO a public profile and air specific issues. Thus it is a means of placing pressure on power-holders. Appearing on the media, especially national television and radio and major newspapers, is seen by many members of the public as a kind of third party endorsement of your position or at least as an indication that your group is a key stake-holder.

You can gain the most benefit from the media if you understand the characteristics and constraints of different kinds of media. The media offer a spectrum of depth and complexity in treating issues. In general, the most shallow and simple form of media is television. Its news programs in particular are great for pictures, brief headlines and showing conflicts between two widely polarised sides. They rarely do justice to complex environmental, social and economic problems such as the mitigation of global warming. So usually the best you can get is a brief, simple 'sound bite'.[14] Try to avoid the temptation to compete in absurdity with the sound bites of the anti-climate action interests. Do the majority of viewers really believe that cutting emissions will take us back to the caves and trees, lighting with candles and travelling on horseback?

A few excellent television current affairs programs and documentaries do offer quality opportunities. Radio in general allows a somewhat greater depth of discussion than television. Again, beware the brief news flashes, unless you have a very simple message to convey. Print media includes newspapers, magazines and newsletters, both hard copy and electronic forms. You can publish letters to the editor, provided they are short, and sometimes opinion pieces and feature articles of up to 1000 words. You can also feed information to journalists via a press release or personal contact.

Developing relationships with journalists over periods of months and years can be an excellent way of gaining media coverage for your group and its cause. You can build trust by ensuring that the information you pass on is accurate, relevant and useful to the journalist and the media outlet. In private background discussions, you should always indicate when you or your group does not have the expertise to answer a particular question and try to suggest a suitable alternative source. Your journalist contacts must in turn be willing and able to understand the issues, which are often complex, and publish a clear and fair account of them.

You could set up a media event to launch the issue or your group's involvement in it and subsequently draw media into a whole range of other tactics. The initial event could be the publication of a study, or an exposure of a targeted politician's failure to implement an election promise, or a demand for information, or a challenge to a new business-as-usual project such as a coal-mine or motorway. To increase your chances of gaining media coverage for a specific event or issue, ensure that:[15]

- the issue is important enough, in the eyes of the media, to warrant coverage;
- the issue and your media statement are clear and unambiguous;
- the issue is relevant to readers and listeners;
- the event is of short duration (a long-lasting issue such as climate change must be broken up into smaller digestible events for the media);
- the event is eye-catching or entertaining;
- if appropriate, some local content or a local personality is involved.

Free local newspapers are widely read and should not be neglected in publicising an event. Your local politicians – from federal, state or local government – will be keen to be included too.

In preparing a media release, the first paragraph is the key. It must grab the journalists' attention and address the four Ws: Who? What? Where? When? Later paragraphs may address Why? and How? Write

the press release on a single page, check that a 14-year-old can understand it and email it to the relevant journalists. Make sure you include
your name and contact details at the end of the release. Follow up the
press release with phone calls offering to clarify it or to provide further
information.[16] An example of a clear, concise and newsworthy media
release is given in box 6.2.

BOX 6.2

Example of a media release

NUCLEAR SAFEGUARDS OFFICE FURTHER DISCREDITS ITSELF

Friends of the Earth

Media Release – 24 October 2008

The Australian Safeguards and Non-proliferation Office (ASNO) is today
bringing itself into further disrepute by accepting an award from a pro-
nuclear lobby group, the Australian Nuclear Association (ANA).

Friends of the Earth nuclear campaigner Dr Jim Green said: "ASNO
is an independent statutory office and should not be accepting an award
from a lobby group. Yesterday, ASNO claimed in Senate Estimates that
the ANA is not a lobby group but an industry group. But the ANA is a
lobby group – its constitution includes the aim of 'promoting ... the use of
nuclear science and technology'. And ASNO is accepting an award from
the industry it purports to regulate.

"The credibility of ASNO is already under serious challenge as its
false and misleading evidence was comprehensively rejected by the
parliamentary treaties committee's recent inquiry into the Howard/Putin
uranium deal. For example, ASNO misled the committee with false
claims that safeguards would 'ensure' that Australian uranium does not
end up in Putin's weapons, but research by Friends of the Earth revealed
that there have been no IAEA inspections in Russia since 2001."

Friends of the Earth has lodged a complaint (available on request)
with the Parliamentary Privileges Committee regarding ASNO's false
'evidence' to the treaties committee.

"ASNO is notorious for peddling false claims, such as its claims that nuclear power does not pose a WMD proliferation risk, that Australia only sells uranium to countries with 'impeccable' non-proliferation credentials, and that all of Australia's uranium is 'fully accounted for'", Dr Green concluded.

More information

- Detailed EnergyScience Coalition critique of ASNO: <www.energyscience.org.au>.

- Friends of the Earth: <www.foe.org.au/anti-nuclear/issues/nfc/mining/safeguards/asno>.

- Dr Jim Green, phone xxx work; xxx home; xxx mobile.

In giving interviews for television and radio, be aware that news programs require concise answers, while current affairs and magazine-type programs will allow you a little more time. In many cases you will find that interviewers have their own agenda, which may differ from yours. There is tension in that situation. Don't try to resolve it by imitating politicians who often blatantly ignore the question and simply repeat their carefully prepared statements. Rather, try to compromise. In responding, first acknowledge the interviewer's question and then turn the answer towards getting out your own message. For example, you could say: 'Yes, that is the electricity generators' argument for compensation for a carbon price, but independent studies present a strong economic case against that position ...'

Nonviolent direct action

Nonviolence is a powerful and just
weapon ... which cuts without
wounding and ennobles the man who
wields it. It is a sword that heals.

Martin Luther King, Jr[17]

183

Nonviolence is not only a set of skills and techniques, but is also a frame of mind and, for some, a philosophy of life.[18]

In the context of climate action, nonviolent actions are undertaken to confront key decision-makers in polluting companies and developments, the financial institutions that fund them, and the governments that are complicit with them. You may use these confrontations to resist a bad government policy or a harmful project, such as a new dirty coal-fired power station or coal mine; to mobilise community opposition; to demonstrate to government the growing community concerns about the issue; or all of the above. Media publicity is a valuable aspect, but is rarely the principal objective of the confrontation, which is to exert pressure on the decision-maker as part of the process of winning a demand.

As protesters, we need to avoid using physical violence for several reasons:

- Most people will not participate in violent actions.
- Violence distracts media attention away from the issue to the violence.
- Violence alienates the wider community and isolates the campaign.
- The state, backed by the military and the police, has by far the greatest capacity to conduct violence. Therefore, any violent action that falls short of a well-resourced popular revolution will inevitably fail.
- Human-induced climate change, together with many other issues of environmental protection and social justice, are the result of violence towards nature and people that is implicit in the actions of power-holders. Meeting violence with violence is morally untenable for solving such problems.
- In practical terms too, violence is untenable because it simply fosters more violence in an endless cycle that is extremely difficult to break. In a quotation attributed to Mahatma Gandhi, 'An eye for eye only ends up making the whole world blind.'

Power-holders are well aware that the use of violence weakens a social movement both internally and in the eyes of the wider community. Therefore they sometimes send agents provocateurs to foment violence that is blamed upon the protest group.[19] For this reason, training in nonviolent action is important for CMOs that are planning nonviolent actions. It can even include methods of nonviolently isolating, surrounding and limiting the actions of a violent minority in a crowd. Nonviolent action training has a wide range of other benefits, as discussed in *Resource Manual for a Living Revolution*.[20] For example:

- it introduces cooperative ways in which people can learn about and change their world;

- it develops skills in conflict resolution and democratic decision-making;

- it allows skills, ideas and organising methods to be developed and tested in practice situations where risks are low and mistakes are less costly;

- it teaches methods of creating trust and solidarity than can be effectively applied to withstand discouragement and repression.

Nonviolent actions include rallies, marches, sit-ins, pickets, naming and shaming, shareholder actions, strikes and boycotts.

Rallies, marches, sit-ins and pickets

Rallies, marches, sit-ins and pickets are ways you can physically, but nonviolently, block or occupy a site, such as a road, public space, gateway, office or whole building.[21] The larger the numbers of protesters, the longer the event can be maintained in the face of police opposition and the more it will attract media attention. Increased numbers will also apply more pressure on targeted decision-makers. To build up the numbers, your organisers should network before the event with a wide range of community groups. In terms of communicating the message clearly with the media and the public, you need to have designated speakers and spokespeople for media interviews, who are well prepared with the agreed message. You should also assemble an appropriate array of banners and signs for media pictures.

Particular businesses and politicians who are clearly working against positive climate action can be targeted by nonviolently picketing or occupying their offices, buildings, gateways. Potential targets include the headquarters of coal-fired aluminium smelting corporations; financial institutions that are funding new power stations, coal-mines and motorways; the manufacturers of 'gas guzzler' motor vehicles; and the offices of ministers who delay or cut funding for renewable energy. When such actions gain media attention, they are a means of naming and shaming the reprobates as well as highlighting a specific issue.

Don't hold pickets, sit-ins and occupations indiscriminately; that would alienate many people and could even strengthen the existing power structure. The whole idea of targeting such actions is to split the power-holders, exposing those who are clearly violating the stated principles and values held by the majority of society and those who are reneging upon election promises.

Naming and shaming

Individuals and organisations that are climate reprobates can be named and shamed in various ways. For example, the British Royal Society published its letter to Exxon Mobil, objecting to the giant oil company's funding of groups that were denying greenhouse science and undermining climate action by governments.[22] Another effective method of naming and shaming is to award a prize for (say) the worst greenhouse polluting company, or the politician who has given the most support to greenhouse pollution, or the land-owner who has cleared the most land of native forest. An amusing well-staged event can expose a serious problem and its perpetrator, while gaining brief but positive media coverage.

For all naming and shaming, it is essential to be sure of the facts and to avoid defamation, for example by casting doubt on someone's motivations.[23] It is entirely proper to point out that a Minister for Energy holds shares in a coal company and is therefore in a position of potential conflict of interest.[24] However, it would be defamatory to suggest that the minister's approval of a new coal-mine was influenced by his/ her shareholdings. If there is any doubt, first seek legal advice before making a public statement targeting an individual or corporation.

Shareholder actions

Another tactic might be to buy a few shares in a greenhouse polluting company. Then go to a shareholders' meeting and ask awkward questions of the chairman of the board. This is a way of challenging the board and other shareholders to be more socially responsible.

Another tactic, described by Saul Alinsky,[25] is to ask shareholders in a company to assign their proxy votes to a CMO. An important part of the process is to lobby and negotiate for proxies with large institutional shareholders in the company. These could be other companies, superannuation funds and in some cases universities. In addition, it may be possible to acquire a large number of individual proxies through internet CMOs such as Avaaz and GetUp, churches, clubs and societies, and environmental NGOs. If enough proxies can be obtained, then the opportunity arises to actually put motions and influence votes at shareholders' meetings.

Withdrawal of deposits

You may wish to discourage a bank or other financial institution from lending money to a project that would be high in greenhouse gas emissions (for example, a new conventional coal-fired power station). In this case your CMO could organise a campaign to encourage depositors to withdraw their savings from the financial institution on a specified date, if it does not withdraw from the proposed loan. Like acquiring proxies, the intention of this tactic is to gain a stronger negotiating position for the CMO. The challenge for the CMO is to use all means of publicity – websites, email lists and media – to mobilise people for the potential withdrawal. Banks only hold typically 10 per cent of deposits in reserve; the remaining 90 per cent is lent to investors. Therefore, a large total withdrawal on a particular date would reduce the bank's ability to make loans and hence its profit margin. A very large total withdrawal could force it to borrow or sell assets to avoid running out of its reserves and defaulting on its obligations. As a bonus, the publicity from this tactic is a valuable part of a public education campaign.

Strikes and boycotts

Strikes and work-to-rule (go-slows) are traditional actions in the struggle to improve working conditions. However, boycotts may be more effective for your group, as they offer targeted actions in the climate crisis.[26]

The two principal types of boycott are consumer boycotts and workers' boycotts. A workers' boycott is a refusal by workers to work with specified materials or tools. It is usually organised by trade unions.

A consumer boycott can be a powerful weapon against businesses that are opposing climate action or against products that are particularly high in greenhouse gas emissions. It can be reinforced with pickets of the businesses or of the manufacturers' offices. Some potential targets are corporations that are funding climate change denier and anti-climate action groups; aluminium products made in mainland Australia (where coal power is used for smelting); all timber that is not certified to be sourced from ecologically sustainable plantations; and poor quality air conditioning, solar hot water and solar PV systems. Actions against products depend on the existence of adequate labelling and certification, so achieving this can be an important intermediate goal of a boycott campaign.

To be effective, all types of boycott have to be well organised, with excellent communications with potential and actual participants. The rise of the internet offers new opportunities in this regard. As in the case of strikes, boycotts can impact on the innocent as well as the target group, so decisions to initiate a boycott should be made responsibly. Strikes and boycotts may not be lawful in some jurisdictions, and penalties could apply. Seek professional legal advice before trying them.

Choosing nonviolent actions

The above tactics are the principal categories of nonviolent actions available to you as a campaigner. A detailed account of nonviolent actions, with numerous examples and categories of action broken down into subcategories, is given in Gene Sharp's classic book, *The Politics of Nonviolent Action.*[27] However, Sharp focuses on nonviolent resistance to invasion by a foreign power and repression by one's own government. Although these can be situations of great suffering, they are in some ways simpler than nonviolent action against a nominally democratic government that is in cahoots with the big greenhouse gas emitters, while not being very oppressive. In this case, much of the population may be complacent about the situation and could easily resent a nonviolent action that inconvenienced it, such as blocking a main road or picketing a widely patronised business or a key product. Indeed, the economic system and laws may give consumers little choice but to use greenhouse-intensive products and services. So, nonviolent actions should be chosen with care and creativity, with an eye to gaining public sympathy and support rather than alienating it.

Legal action

Before taking legal action a CMO needs two things: suitable legislation and good lawyers willing to undertake a case in the public interest either pro bono (at no cost) or at low cost. Given these conditions, court cases may be a useful tool in some jurisdictions.

An example of a successful case was the action taken by the Sierra Club before the US Environmental Protection Agency's Environmental Appeals Board in 2008. The board overturned the air pollution permit that had been previously granted for a waste coal-fired power station proposed by Deseret Power, on the grounds that it failed to require any controls on carbon dioxide pollution.[28]

Another successful court case was one in which environmental activists were defendants. Six Greenpeace activists painted the British prime minister's name on the chimneys of the Kingsnorth power station, the site of a proposed new giant dirty coal-fired power station. The action caused damage estimated at £35 000 and the activists were brought to trial. Jurors found that the six had a lawful excuse to damage property to prevent even greater damage from climate change. In the UK, the defence of 'lawful damage' and the *Criminal Damage Act 1971* allows damage to be caused to property to prevent even greater damage, such as breaking down the door to a burning house to fight a fire.[29]

Incidentally, according to Greenpeace UK, the British Government is planning to refer the case to the Court of Appeal in order to take the decision out of the hands of juries and remove the defence of 'lawful damage' from activists.[30] If true, this could be seen as a weakening of the democratic process in the interests of greenhouse polluters. The struggle against global warming is also a struggle for a more democratic system of governance.

Successful legal actions by community groups against governments and large corporations are rare. All the advantages are with those who have money and power. Before embarking on legal action against a government, consider that governments make the laws and, if they lose a court case, they have been known to change the law to ensure future legal victories. For example:

- In Canberra (Australia), a group of citizens successfully challenged in the courts the construction of a huge telecommunications tower in the Black Mountain nature reserve. Then the government simply changed the law and constructed the tower.[31]

- In Victoria (Australia), when the operators of Australia's dirtiest coal-fired power station, Hazelwood, sought consent to develop a new open-cut coal-mine and so extend the life of the power station, CMOs took successful legal action to require the greenhouse impact to be considered. The state government then granted consent to the coal-mine.

Before embarking on legal action, a CMO should obtain good legal advice, considering whether any laws and precedents can be used, and carefully calculate the likely costs. Even if a legal practice agrees to act pro bono for the CMO, the costs of bringing expert witnesses from overseas and the opportunity costs of tying up campaigners in long, complicated legal procedures should be evaluated.

Setting up alternatives

Community projects to reduce greenhouse gas emissions:

- demonstrate to power-holders a strong community commitment to change;
- provide public education, precedents and models of development that other communities can follow;
- sometimes demonstrate new technology;
- sometimes provide economic savings; and
- generally give solidarity and a sense of power to a community.

They also provide opportunities for local politicians and businesses to become identified with the project and give it (and them) more status as a result.

Projects may involve, for example, a community housing development based on solar efficient social housing; bulk purchase of insulation, solar hot water or solar photovoltaic systems for residents of a region; or a wind farm or solar power station developed and owned by the local community.

The modern wind power industry arose in Denmark in the 1970s and '80s on the foundations of community ownership of wind turbines. This was facilitated by government funding of research and development, a feed-in tariff and legislation that made community ownership of wind turbines administratively simple. Outside Denmark, progress in community-owned renewable energy projects has been slow. In the UK, the Westmill Co-op Wind Farm commenced operation in 2008 on the property of a local organic farmer. It has five turbines, each rated at 1.3 MW, and is 100 per cent community owned.[32] In

Australia, with unsympathetic governments and no feed-in tariffs for renewable energy power stations, it has been difficult to establish community-owned renewable energy power stations. By coincidence, the first proposal for a community-owned wind farm was made in a small town called Denmark in Western Australia. It would have comprised two turbines each rated at 900 kW. Unfortunately local opposition seems to have stalled this initiative. In 2009, a community-owned wind farm is being developed by the Hepburn Renewable Energy Association in central Victoria. It is planning to construct two wind turbines each rated at 2 MW.[33] Meanwhile, another group, Clean Energy for Eternity, is in the early stages of developing a proposal for a community-owned solar power station.[34]

Choosing tactics

Soft or hard words?

It is better for a decision-maker to agree voluntarily than under pressure. Then the decision is likely to be implemented faster and in greater depth, with a more substantial support structure. Therefore, if there is a reasonable chance that official channels will be responsive, a CMO should go through them first, by making submissions to official enquiries and hearings, writing letters to decision-makers, face-to-face lobbying, and any available legal remedies. Occasionally the decision-maker will agree or at least propose an acceptable compromise. However, as recognised by stage 2 of the Movement Action Plan, in the majority of cases all these activities simply serve to demonstrate the failure of official processes and institutions.

This failure may occur even when the decision-maker may be personally sympathetic to your case. In their official position, they must reflect government or company policy and this is often determined by powerful vested interests. These interests are powerful because they are rich, pay large amounts of taxes (although they are expert in minimising them), make large political donations and can gain extensive media coverage for their positions. So, the CMO must then move to place pressure on the decision-makers, drawing upon a wide range of nonviolent action techniques, media events, and events that educate the public and stimulate their active participation.

It must be emphasised that lobbying and other negotiations, nonviolent direct action, door-knocking and doing media interviews should not be conducted without training and rehearsals under the guidance of experienced mentors. Even writing opinion-pieces and letter-writing will benefit from feedback from other members of the group. Please be aware that some tactics – for example, strikes and boycotts – are illegal in some countries, even those that purport to be democracies. Make sure that all members of your group understand the legal consequences of each action.

Tactics used by different types of CMOs

In general, the large national and state generalist NGOs tend to place their principal efforts on lobbying federal and state governments, building alliances with sympathetic businesses and trade unions, commissioning reports and obtaining media coverage on national and state scales.

The smaller climate action groups are dedicated predominantly to local climate action, including guidance to members on reducing their own greenhouse gas emissions, gaining local media coverage, lobbying local governments and local members of parliament, organising bulk purchases of solar hot water and solar PV systems and setting up local demonstrations of renewable energy.

The business-NGO alliances that seek strong action to reduce emissions, mainly commission reports on greenhouse science and the economics of greenhouse mitigation. These reports are generally publicised by the large generalist environmental NGOs which in some cases are the conveners of the business groups.

Surprisingly, peak business organisations for renewable and energy efficiency have had little presence in the Australian media lately. The former Business Council for Sustainable Energy was very active, but after it amalgamated with the Australian Wind Energy Association in

2007 to form the Clean Energy Council, it lost public prominence. After the May 2008 budget that delayed implementation of most of the election promises of Rudd Labor for renewable energy (see chapter 2), the Clean Energy Council even issued a press release praising the budget:

> The Clean Energy Council, the peak national industry body for the clean energy and energy efficiency sectors, said the Rudd government delivered on its climate change election promises in their first federal budget.[35]

An environmental journalist has suggested that fossil fuel businesses have taken over the Clean Energy Council. Internal tensions within this organisation have erupted into a court case.[36]

Meanwhile, business groups opposed to positive action on climate change commission their own reports, obtain extensive media coverage and lobby governments heavily.[37]

Trade unions are mainly informing their members about the key issues, especially the implications of greenhouse mitigation policies for employment and for geographic regions. Those that favour climate action are developing and putting forward policies for a 'just transition' to a low-carbon future.

The diversity of CMOs and their tactics is one of the strategic strengths of the movement, enabling different groups to appeal to different segments of the population. Therefore, movement organisers would be making a mistake if they pressured all CMOs to have identical policies and tactics. That would be a sure recipe for marginalising the movement.

7

THE STRUGGLE FOR A
SUSTAINABLE FUTURE

Nonviolent social movements have toppled colonial powers and dictators, gained equal rights for African Americans, won votes for women, ended official support for slavery, transformed poverty-stricken regions, developed trade unions to protect the rights of workers, ended the testing of nuclear weapons in the atmosphere, and created and protected magnificent national parks.

Now the Earth and its peoples are facing a challenge that is possibly even greater than all of these, a climate emergency. Positive feedback processes are amplifying the emission of greenhouse gases and global climate change. Unless these processes are terminated quickly, they will drive the Earth inevitably and irreversibly into a much warmer and hostile climate characterised by extremes of heatwave, drought, bushfire, flood, hurricane, rising sea-level and the loss of many species.

Thanks to climate scientists who are elucidating the growing evidence of a changing climate, many people are aware that a serious global problem exists. But our governments are not yet treating it as an emergency, instead seeing the climate crisis as just one of several political issues on their lists, but not nearly as important as the global economic crisis. They appear to believe that they can trade off the science against the politics. Conventional dirty coal-fired power stations and motorways are still being built. Oil is still being extracted from tar sands and there are schemes to produce oil from coal, despite the very high greenhouse gas emissions of these processes. A number of governments have set targets for reducing greenhouse gas emissions by 2050, but very few outside Europe have set targets to reduce emissions significantly below the 1990 level by 2020.

The principal barriers to change are the powerful vested interests that strive to maintain business-as-usual at all costs to the rest of society and the planet, and the associated economic system that fosters endless growth in the consumption of energy, materials and land, even though this is unsustainable.

The principal hope of changing this situation is the growing social movement for climate action. This movement is international, national and local. It has already had a significant impact on the policies of several national governments in Europe and the EU as a whole, and on the policies of many state and municipal governments in the USA. It has influenced President Barack Obama to dramatically increase funding for renewable energy and possibly to change the position of the US federal government on climate action on the international scene.

But, it can also be argued that the climate action movement has reached an impasse; so far no country is treating climate change as an emergency. Within the conceptual framework of the Movement Action Plan (see chapter 5), a major shock is needed to move the campaign from stage 3 'Ripening conditions' to stage 4 'Take-off'. Let's consider three ways in which this could occur.

Firstly, a sudden change in the climate and its impacts may occur, leading to widespread public concern on a global scale. A possible example would be the collapse of part of the West Antarctic Ice Sheet, causing a very large quantity of ice to slide from land into the sea, possibly resulting in a sea-level rise of one metre or more within a period of weeks. The climate movement must prepare for such an event, if and when it occurs, using it as a trigger for emergency action.

Secondly, a political scandal, such as the revelation of a conspiracy or collusion between greenhouse gas polluters and a government to hold back renewable energy, could possibly lead to a change of policy in the country concerned. Such a scandal did occur in Australia under the Howard Government, as described in chapter 2, but few were aware of it. As a climate campaigner you might wish to initiate investigations of potential collusion, seeking evidence from whistleblowers to put before the public via the media.

Thirdly, a long hard struggle at the grass-roots level is needed to build up the numbers and political influence of the climate movement, to expose the role of the vested interests in delaying and undermining the solutions (chapter 2), and to create a climate of public opinion that will no longer be fobbed off with token gestures and 'spin' from government.

Realistically, the climate action movement cannot afford to wait for the first two events to occur. Strategically those of us in the movement have no choice but to take the third pathway, building upon our strengths: our large numbers, our diversity and our integrity. We must further increase our numbers and organise them. We must educate the public and the media, and exert the pressure of our organised numbers upon the power-holders in government, business, trade unions and professional organisations. We must convince more and more people in these sectors to support the movement. We must make good use of our large portfolio of tactics comprising lobbying and negotiating with power-holders, networking, educational activities, media, nonviolent direct action, legal action, and setting up alternatives.

Two groups in particular need to play much more active roles in the climate action movement. Scientists have alerted us to the climate crisis, but only a few have spoken out publicly to refute deniers and to demand effective mitigation policies from governments. James Hansen in the USA, one of the world's leading climate scientists, is the most notable international public commentator. Much of the material disseminated by the deniers of human-induced climate change is bogus, although it is marketed as credible science in the same way that so-called 'creation science' and 'intelligent design' are marketed as scientifically based. Scientists have been quite active in refuting the notion that these fundamentalist religious doctrines are science, yet many scientists have failed to challenge the spurious nature of opposition to climate science and action.

Religious leaders have special responsibilities to speak out to their congregations and the public at large on the moral issues inherent in climate change and mitigation. Some are already doing this, but more are needed. Faith groups have a much greater potential role in the climate action movement. Indeed, all people who are concerned about ethical behaviour must speak up.

The proponents of business-as-usual can see the writing on the wall for greenhouse polluting industries. As a consequence, some are pushing with desperation to build new dirty coal-fired power stations, oil from coal and shale projects, motorways, smelters, etc, before these projects are widely recognised as monstrosities and banned. The climate action movement must shut down these leftovers from the smokestack age. At the same time it must work constructively to create an ecologically, economically and socially sustainable society. This is the struggle for the future of the Earth and humanity.

RIDING ON TRUST AND RIPE CONDITIONS IN THE UK

Nina Hall, Graduate School of the Environment,
Macquarie University

In the United Kingdom (UK), three significant CMOs are Friends of the Earth England, Wales and Northern Ireland (FoE-EWNI), Greenpeace UK, and WWF-UK (formerly the Worldwide Fund for Nature UK). These are all organisations linked to international networks and offices. Each of these CMOs has undertaken a different campaigning approach. FoE-EWNI regards itself as a campaigning organisation whose job is to raise the standards that others, such as the Government, are charged to implement, Greenpeace UK's role as a protest organisation works to exploit media attention to pressure the Government and corporations, and WWF-UK maintains ongoing involvement and access with the Government. These UK CMOs have campaigned on climate change using strategies that include, among others, a focus on election campaigns, attempts to influence policy-making, and efforts to build grass-roots public action.

Election campaigns

In the lead-up to the 2005 UK elections, WWF-UK selected several newsworthy marginal seats that had experienced adverse climate change-related impacts, including Lewes in south-east England. Lewes was affected by a hurricane in October 1987, droughts in 1995, and severe flooding in October 2000. WWF-UK campaigners determined that climate change issues would resonate more with Lewes residents than in some other electorates and therefore chose to integrate climate change demands into the political agenda of voters. One of the WWF-UK activities involved a hustings event on climate change policies where all local candidates seeking election were asked questions by the constituents. WWF-UK placed radio and newspaper advertisements, and projected a campaign film about climate change impacts onto the side of the House of Commons in London. Following the 2005 UK election which resulted in the Blair Government retaining power, UK CMOs continued to encourage constituents to discuss climate change with their political representatives. FoE-EWNI organised regular public events featuring MPs to pressure these MPs to demonstrate that the constituents wanted action on climate change.

Involvement in policy-making

Draft legislation that later became the UK Climate Change Bill was initiated by FoE-EWNI and WWF-UK, acting in a broader collaboration as 'The Big Ask' campaign. This campaign encouraged people to contact their local MPs to support the early form of the Bill, known as an Early Day Motion. Citizens were mobilised through local chapters of their organisations, including through over 200 local FoE-EWNI groups around the UK. FoE-EWNI organised rock concerts and encouraged concert-goers to send an SMS message to politicians asking them to support the Motion, distributed information at Summer festivals, ran a cinema advertisement, and used viral marketing[1] to circulate cartoons and jokes by email with a link to the campaign website. Lobbying by approximately 130 000 members of the voting public resulted in support for the Motion from 400 MPs and the announcement of the UK Government's endorsement of the draft Climate Change Bill in March 2007. This Bill mandates greenhouse gas emission reductions of 26 to 34 percent by 2020 and 80 percent by 2050 on 1990 levels. The Climate Change Bill was passed by the UK Parliament and received Royal Assent in November 2008.

This Bill ensures that the UK is the first country to propose legislation setting binding limits on greenhouse gas emissions beyond its commitments under the Kyoto Protocol.

Popular grass-roots awareness and community action

FoE-EWNI, WWF-UK and Greenpeace UK are all members of Stop Climate Chaos (SCC), an umbrella network of UK environmental groups, trade unions, faith organisations and womens' groups working on climate change. SCC was established in 2005 intentionally to move climate change from a marginal issue to a collaborative campaign upon which CMOs from a variety of sectors could unify their campaign message. Previous analysis by the member groups of SCC determined that public campaigning on the scale and commitment necessary to provide Government with a mandate for climate action was lacking in many CMO campaigns. SCC was established to act as a conduit for such public pressure.

Working collaboratively with and beside its members, SCC began their activities by raising community awareness of climate change before encouraging people into taking action. The member CMOs structured their awareness-raising messages positively to increase community support and attention, despite the sobering subject of climate change. For example, the FoE-EWNI awareness-raising materials and website intentionally maintained an upbeat tone to motivate people. FoE-EWNI supported the formation of local climate action groups and supplied them with a regular newsletter of suggested activities. Additionally, FoE-EWNI increased awareness of the issue through a national 'Shout About Climate Change' week of school activities, where activity and information kits were sent to approximately 1700 schoolteachers and youth workers.

The next stage of the community-focused campaigns was to translate the awareness into action. UK CMOs sought via SCC's diverse coalition to activate participants from a cross-section of society to ensure that calls to politicians for action on climate change came from broader civil society, making it harder to ignore. One of SCC's activities was the I Count campaign, with the slogan 'Together we are irresistible – together we count'. This campaign brought together 25 000 people, including celebrities and public figures, for a carnival-like rally from the US embassy to Trafalgar Square, London, for the International Day of Action on Climate Change on 4 November 2006. This was the eve of the United Nations Framework Convention on Climate Change conference in Nairobi.

Supportive political context

The achievements by the UK CMOs stem from conducive political and policy conditions provided by the UK Government over the last decade. There was very strong political posturing on climate change by the Blair Government (1997–2007), both domestically and within the European Union. Until the decision to build the new conventional coal-fired power station at Kingsnorth, the fossil fuel lobby did not appear to have predominant influence. The Kingsnorth development was cited by some as proof that the Business Secretary 'bought into the arguments of the electricity companies that coal can, in fact, be "clean" … we need to be very wary of such "greenwash"'.[2]

More positively, CMOs are highly regarded by the UK Government and community and have millions of supporters from whom to draw their main financial support. UK CMOs appear to have contributed to substantial climate change policy modifications through this strong credibility they hold with the Government.

The information in this appendix is extracted from its author's following papers: Hall, N & Taplin, R (2007) Solar festivals and climate bills: Comparing NGO climate change campaigns in the UK and Australia, *Voluntas* 18(4): 317–38; Hall, N & Taplin, R (2009, in press) Empowerment of individuals and realisation of community agency: Applying Action Research to climate change responses in Australia, *Action Research*.

BURGEONING GRASS-ROOTS
POWER IN THE USA

*Nina Hall, Graduate School of the Environment,
Macquarie University*

There is a broad variety of CMOs in the US that all seek, through
a variety of campaign activities, to influence US federal and state
climate decisions and policy development. Six CMOs from among
this diversity include:

- *Sierra Club* was established in 1869 and proclaims that it is
  the most influential grass-roots environmental NGO in the
  US with 750 000 members. It expanded its original wilderness
  preservation mission into wider environmental issues, including
  climate change.

- *Union of Concerned Scientists* has over 150 000 members,
  forty percent of whom are scientists. Its mission is to combine
  scientific analysis, policy development and citizen advocacy
  for practical environmental solutions on a number of issues,
  including climate change.

- *Greenpeace USA* has gained high popularity and effectiveness
  from using 'aggressive' local activism and 'flamboyant' media
  events to mobilise public support on environmental issues,
  including climate change.

- *Bluewater Network* has a strong focus on legislative change, and
  employs both professional lobbyists and litigation lawyers.

- *Vote Solar Initiative* was established at municipal level after a successful vote for renewable energy in San Francisco, and now works at municipal, state and federal levels.
- *Apollo Alliance* was formed between labor unions, economic, business, social justice and faith-based organisations, philanthropic foundations and environmental groups.

The above CMOs have played an active role in lobbying for, and publicly supporting, the governmental decisions to develop the climate-related policies, have publicly endorsed political candidates, have tailored campaign messages to be delivered by specific messengers, and have built alliances to broaden their message.

A crucial role in the development of California's Assembly Bill 1493 that sought to reduce automobile emissions was played by the Bluewater Network. They conceived, drafted and championed the Bill and worked closely with Assembly Member Fran Pavley to introduce it. Other CMOs joined Bluewater's efforts later in the campaign to build the call for this Bill. A similar legislation-focused campaign was undertaken by Vote Solar. It engaged with governments and utilities to increase the 'net metering cap', a solar power subsidy that allows Californian households to receive full retail value for the eligible solar and wind-generated electricity they produce and export to the electricity grid. Vote Solar's lobbying efforts for this policy involved facilitating 50 000 emails from the general public to political decision-makers. The resulting net metering cap was adopted as part of the 'one million solar roofs' program and legislated in Senate Bill 1.

In 2006, the Sierra Club assessed the climate and other environmental policy pledges of the candidates for the election of Californian Governor, and publicly endorsed the Democrat candidate over Republican incumbent, Governor Schwarzenegger. This work was undertaken by the Club's political action committee, which also directs political contributions. The Sierra Club mobilises its members during elections. This includes encouraging its large supporter base to actively lobby their political representatives for specific efforts, such as voting in 2005 for California's Assembly Bill 32 to reduce greenhouse gas emissions to 1990 levels by 2020.

The Union of Concerned Scientists' 'Sound Science Initiative' trains its scientist members in media and communication skills. On climate change, these members communicate the findings of Union-commissioned climate-related research reports at press conferences, where Union campaigners then propose policy recommendations. With this approach, climate campaign messages resonate with specific audiences and the media while maintaining the scientists' impartiality.

To broaden the political influence of their campaign messages, some CMOs have built coalitions with diverse organisations. The Apollo Alliance formed as a coalition of environmental, labor, business, and community allies who 'share a common vision for the future and a common set of values' and that seeks to align economic development with 'strong action on global warming'. Similarly, the Sierra Club published a report as a 'blue-green alliance' with workers and unions that described the potential for 1.4 million new jobs to be created if renewable energy policies were implemented.

US CMO campaigners have stated that fossil fuel and related industry interests have had a dominant influence on federal climate policy decisions, particularly under the Bush Administration. Strong business influence in political decision-making is also enabled by the focus of federal politicians on political campaign fundraising. The CMO campaigns have responded to this situation by undertaking state-level intervention with campaigns focused on influencing state policies, including the above-mentioned campaigns in California. Campaigns and initiatives at a municipal and state level can 'create the interest and involvement that presage much-needed political change' that can flow upwards to a federal level.

Such community- and state-based campaign action has had significant outcomes, especially on the development of new coal-fired power stations. In 2007, over 25 proposals for new stations were cancelled. These cancellations were cited as the result of 'strong public opposition, combined with uncertainty over the future costs of complying with carbon dioxide emission caps and concerns about global warming's environmental and economic impacts', as well as the new greenhouse regulations in California and other states. Legal action and grass-roots protests and lobbying by members of CMOs, including the Sierra Club and university students, contributed to the closure and replacement of two aging coal-fired power stations in Wisconsin with cleaner systems. More recently, 17 proposed coal-fired power stations were 'cancelled, abandoned, or put on hold in the United States' in 2008. These developments suggest that the CMOs, incorporating significant agitation from the grass-roots level, have been influential in the future energy decisions in US states.

This information is extracted from the author's following work: Hall, N (2008) *Obstacles and Opportunities: An analysis of climate change campaigns by Australian NGOs*, PhD Thesis, Macquarie University, Sydney. References therein.

NOTES

Unless otherwise indicated, all websites listed here were accessed in March 2009.

Chapter 1: Threat and hope

1   Diamond, J (2005) *Collapse: How societies choose to fail or survive*, Allen Lane, Melbourne, chapter 2.
2   Intergovernmental Panel on Climate Change (IPCC) (2007a) *Climate Change 2007: The physical science basis*, Summary for Policymakers, Intergovernmental Panel on Climate Change, p 10, <http://www.ipcc.ch>.
3   Hansen, J, Sato, M, Ruedy, R & Lo, K (2009) 2008 Global Surface Temperature in GISS Analysis, 13 January, <http://www.columbia.edu/~jeh1/mailings/2009/20090113_Temperature.pdf>.
4   Stroeve, J, Holland, MM, Meier, W, Scambos, T & Serreze, M (2007) Arctic sea ice decline: faster than forecast, *Geophysical Research Letters*, 34, L09501, coupled with interview of co-author Ted Scambos in Zabarenko, D (2007) Arctic ice-cap melting 30 years ahead of forecast, *Reuters*, 1 May, <http://www.reuters.com/article/scienceNews/idUSN0122477020070501>.
5   IPCC (2007a).
6   Hansen, J et al (2008) Target atmospheric CO_2: Where should humanity aim? *Open Atmospheric Science Journal*, 2: 217–31; Pittock, AB (2009) *Climate Change: The science, impacts and solutions*, CSIRO Publishing and Earthscan, Collingwood.
7   Stern, N (2006) *Stern Review: The economics of climate change*, October, <http://www.occ.gov.uk/activities/stern.htm>. Actually Stern considered conservative global warming scenarios. He did not consider the positive feedback processes (discussed within this chapter) that are beginning to amplify global warming and could lead to major irreversible changes and hence much greater economic impacts.
8   Ironically, the biggest aftershock of nuclear war involving even a tiny percentage of current arsenals is likely to be *temporary* global cooling or 'nuclear winter' produced by the smoke, dust and ash lofted into the atmosphere by fires and explosions. This could dramatically cut world food production for several growing seasons. See Toon, OB, Turco, RP, Robock, C et al (2007) Atmospheric effects and societal consequences of regional scale nuclear conflicts and acts of individual nuclear terrorism, *Atmospheric Chemistry & Physics* 7: 1973–2002; Robock, A, Oman, L, Stenchikov, GL et al (2007) Climatic consequences of regional nuclear conflicts, *Atmospheric Chemistry & Physics* 7: 2003–12.

9   Diesendorf, M (2007a) *Greenhouse Solutions with Sustainable Energy*, UNSW Press, Sydney.

10 Diesendorf (2007a).

11 Saddler, H, Diesendorf, M & Denniss, R (2007) Clean energy scenarios for Australia, *Energy Policy* 35(2): 1245–56. This paper shows that Australia, the biggest per capita emitter of the industrialised world, could halve its 2001 CO_2 emissions from the stationary energy sector by 2040, based on existing commercially available technologies. By adding large contributions from solar and hot rock geothermal power, close to 100 per cent reduction in emissions could be achieved. Teske, S & Vincent, J (2008) *Energy [R]evolution: A sustainable Australia energy outlook*, Greenpeace Australia Pacific and European Renewable Energy Council, <http://www.energyblueprint.info>.

12 For the USA, see Makhijani, A (2007) *Carbon-Free and Nuclear-Free: A roadmap for US energy policy*, IEER Press, Maryland and RDR Books, Muskegon MI.

13 Although Australia does not have nuclear power, it is the world's biggest uranium producer and is expanding that industry.

14 This is now recognised to be a misquotation of the following statement of Charles E Wilson, president of General Motors from 1941–53: '… for years I thought what was good for the country was good for General Motors, and vice versa'.

15 Moyer, B, with McAllister, J, Finley, ML & Soifer, S (2001) *Doing Democracy: The MAP model for organizing social movements*, New Society Publishers, Gabriola Island BC, Canada, p 10.

16 From US President Abraham Lincoln's Gettysburg Address (1863).

17 Beder, S (2000) Global Spin: The corporate assault on environmentalism, revised ed, Scribe, Melbourne.

18 There are of course huge variations within this average, between the poles and equator.

19 IPCC (2007a).

20 Schmidt, G (2007) The CO_2 problem in 6 easy steps, <http://www.realclimate.org/index.php/archives/2007/08/the-co2-problem-in-6-easy-steps>; Weart, S (2008) The discovery of global warming, American Institute of Physics <http://www.aip.org/history/climate/co2.htm>; many physics textbooks, chapters on heat and thermodynamics.

21 Pittock (2009).

22 Hansen, J (2008) Tell Barack Obama the truth – the whole truth, letter to the President-elect, revised 29 December, <http://www.columbia.edu/~jeh1>.

23 IPCC (2007b) Climate Change 2007: Impacts, adaptation and vulnerability, p 397, <http://www.ipcc.ch>.

24 Karoly, D (2009) Bushfires and extreme heat in south-east Australia, *Real Climate*, 16 February, <http://www.realclimate.org/index.php/archives/2009/02/bushfires-and-climate/langswitch_lang/sw more-654>.

25 IPCC (2001) *Climate Change 2001: Synthesis report*, Summary for Policymakers, An assessment of the Intergovernmental Panel on Climate Change, p 14, <http://www.ipcc.ch>.

26 IPCC (2007c) *Climate Change 2007: Synthesis report*, An assessment of the

Intergovernmental Panel on Climate Change, p 65, <http://www.ipcc.ch>.

27 National Snow and Ice Data Center, USA (2008) *Arctic sea ice down to second-lowest extent; likely record-low volume*, Press release, 2 October, <http://nsidc.org/news/press/20081002_seaice_pressrelease.html>.

28 Pittock (2009) chapter 5.

29 Pittock (2009) chapter 5.

30 IPCC (2007a).

31 Park, G-H, Lee, K & Tishchenko, P (2008) Sudden, considerable reduction in recent uptake of anthropogenic CO_2 by the East/Japan Sea, *Geophysical Research Letters* 35: L23611, DOI: 10.1029/2008GL036118.

32 Ruttimann, J (2006) Oceanography: sick seas, *Nature*, 442: 978–80.

33 Hansen, J, Sato, M, Kharecha, P et al (2007) Climate change and trace gases, *Philosophic Transactions of the Royal Society A* 365: 1925–54. This paper draws upon paleoclimate data to investigate primarily the effect of 'albedo flip' (the change in Earth's reflection of sunlight from snow and ice) on climate. There are few papers on the effects of other positive feedbacks, apart from water vapour.

34 Even in Australia, a country that has large resource industries, mining is only 5 per cent of gross domestic product (GDP). Adding processing and metals production could lift this to about 10 per cent. However, most Australians are given the incorrect notion by politicians and the media that these industries contribute the majority of GDP. They are important for export income, but they not necessarily an essential feature of the Australian economy. See Pearse, G (2009) Quarry vision: Coal, climate change and the end of the resources boom, *Quarterly Essay* 33.

35 Milne, C (Senator) (2008) Some are more equal than others, *ABC News*, 16 December, <http://www.abc.net.au/news/stories/2008/12/16/2447343.htm>.

36 Global Commons Institute <http://www.gci.org.uk>.

Chapter 2: Greenhouse Mafia and their fallacies

1 Extending the perspective beyond the energy sector reveals that Australia's large population of cattle and sheep is also a significant factor in its very high per capita CO_2-equivalent emissions.

2 Sourcewatch <http://www.sourcewatch.org/index.php?title=Global_Climate_Coalition>.

3 Sourcewatch <http://www.sourcewatch.org>.

4 Revkin, AC (2005) Bush aide softened greenhouse gas links to global warming, 8 June, <http://www.nytimes.com/2005/06/08/politics/08climate.html>.

5 Australian Broadcasting Corporation (ABC) (2004) Leaked documents reveal fossil fuel influence, *PM*, 7 September, <http://www.abc.net.au/pm/content/2004/S1194166.htm>; Hamilton, C (2007) *Scorcher: The dirty politics of climate change*, Black Inc Agenda, Melbourne, pp 10–12.

6 Pearse, G (2007) *High and Dry*, Viking, Melbourne.

7 Diesendorf, M (2007a) *Greenhouse Solutions with Sustainable Energy*, UNSW Press, Sydney, pp 107–09.

8 Climate Institute (2007) *Climate Institute: Marginal electorates exit poll*, conducted by Australian Research Group Pty Ltd, November, <http://www.

climateinstitute.org.au/images/stories/exitpoll.pdf>.

9 The intense lobbying by the Greenhouse Mafia is reflected in ABC (2009) Heat on the hill, *Four Corners*, 9 March, <http://www.abc.net.au/4corners/content/2008/s2511380.htm>.

10 Australian Labor (2007) *Energy Innovation; Renewable Energy Fund; National Clean Coal Initiative; Clean Business Australia*, Fact sheets, <http://www.kevin07.com.au>, accessed November 2007 (no longer accessible); Rudd, K, Garrett, P & Evans, C (2007) *Labor's 2020 Target for a Renewable Energy Future*, Election 2007 Policy Document, October, <http://www.alp.org.au/policy/2007policydocs.php>; Evans, C (2007) *Securing a Sustainable Energy Supply for Australia's Future*, Election 07, <http://www.alp.org.au/policy/2007policydocs.php>.

11 Australian Government, Budget 2008–09, <http://www.budget.gov.au>.

12 Australian Government, Department of Resources, Energy and Tourism (2008) Energy programs, 19 December, <http://www.ret.gov.au/energy/energy%20programs/Pages/EnergyPrograms.aspx>.

13 Australian Government, Department of Resources, Energy and Tourism (2008).

14 The principal allocation from the $150 million Energy Innovation Fund, made after the budget was announced, was $50 million for geothermal drilling, which is not even research. Subsequently this allocation was transferred to the $500 million Renewable Energy Fund. No portion of the Energy Innovation Fund was allocated for actual research in 2008.

15 ABC (2008) Wind/solar industries stall, *7.30 Report*, 8 December, <http://www.abc.net.au/7.30/content/2008/s2440907.htm>.

16 COAG (2008) *Design Options for the Expanded National Renewable Energy Target Scheme*, COAG Working Group on Climate Change and Water, <http://www.climatechange.gov.au/renewabletarget/consultation/pubs/ret-designoptions.pdf>.

17 Australian Government (2008) *Carbon Pollution Reduction Scheme: Australia's low pollution future*, White Paper, Canberra, <http://www.climatechange.gov.au>.

18 Murphy, K (2008) Climate change: Act now, *The Age*, 21 February, <http://www.theage.com.au/news/environment/climate-change-act-now/2008/02/21/1203467281202.html>.

19 Australian Government (2008).

20 Garnaut, R (2008) *The Garnaut Climate Change Review: Final report*, Cambridge University Press, Cambridge, <http://www.garnautreview.org.au/pdf/Garnaut_prelims.pdf>.

21 Climate Institute (2008) *Government's climate change credentials slump*, conducted by Auspoll, 23 October, search <http://www.climateinstitute.org.au>.

22 For a concise history of nuclear power in the UK, see Aldred, J (2008) Timeline: Nuclear power in the United Kingdom, *The Guardian*, 10 January, <http://www.guardian.co.uk/environment/2008/jan/10/nuclearpower.energy>.

23 Greenpeace UK (2008) *Whitehall emails reveal government climate policy being dictated by German utility giant*, Media release, 31 January, <http://greenpeace.org.uk/media/press-releases/government-climate-policy-dictated-by-german-utility-giant-20080131>; see also linked documents obtained under Freedom of

Information. More on Kingsnorth in chapter 6.

24 Department of Trade & Industry (2003) *Our Energy Future: Creating a low-carbon economy*, Section 4.68, search for Publication URN 03/660 at <http://berr.ecgroup.net/Search.aspx>.

25 UK Department of Business (2008) *Meeting the Energy Challenge: A White Paper on nuclear power*, <http://www.berr.gov.uk/whatwedo/energy/sources/nuclear/whitepaper/page42765.html>.

26 Adam, D (2006) Royal Society tells Exxon: Stop funding climate change denial, *The Guardian*, 20 September, <http://www.guardian.co.uk/environment/2006/sep/20/oilandpetrol.business>.

27 Gore, A (2006) *An Inconvenient Truth*, St Martins Press, New York. Also film and video.

28 Real Climate (a website by climate scientists) <http://www.realclimate.org>; Schneider, S <http://stephenschneider.stanford.edu>, and in particular the pages on contrarians: <http://stephenschneider.stanford.edu/Climate/Climate_Science/CliSciFrameset.html>.

29 Anderson, RC (1998) *Mid-Course Correction: Towards a sustainable enterprise: The Interface Model*, The Peregrinzilla Press, Atlanta GA; Anderson, RC (2005) *On Responsibility in the Private Sector*, Second International Conference on Gross National Happiness, Rethinking Development, Local Pathways to Global Wellbeing, St Francis Xavier University, Antigonish, Nova Scotia, Canada, 20–24 June, <http://www.gpiatlantic.org/conference/proceedings/anderson.htm>.

30 Keys, K Jr (nd) *The Hundredth Monkey*, <http://www.wowzone.com/100th.htm>.

31 Oakeshott, R (1978) *The Case for Workers' Co-ops*, Routlege & Kegan Paul, London; Thomas, H & Logan, C (1982) *Mondragon: An economic analysis*, George Allen & Unwin, London; Whyte, WF & Whyte, KK (1991) *Making Mondragon: The growth and dynamics of the worker cooperative complex*, Cornell University Press, Ithaca NY.

32 Korten, DC (2001) *When Corporations Rule the World*, 2nd ed, Kumarian Press, Bloomfield CT, USA.

33 At the end of climate talks in Poland, Dr Jiahua Pan, from the Chinese Experts Committee for Climate Change, said that Australia would be acting as though it considered itself a poor nation if it set a maximum target of a 15 per cent cut. Reported in Morton, A & Nicholson, B (2008) Australia called to act on climate, *The Age*, 8 December, <http://www.theage.com.au/national/australia-called-to-act-on-climate-20081207-6t8o.html?page=-1>.

34 Johns, G (2008) Carbon tax is just tilting at windmills, *The Australian*, 27 October. See also the polemic: Lawson, N (2009) *An Appeal to Reason: A cool look at global warming*, Duckworth, London.

35 Clemons, EK & Schimmelbusch, H (2007) *The Environmental Prisoners' Dilemma or We're All in This Together: Can I trust you to figure it out?* <http://opim.wharton.upenn.edu/~clemons/blogs/prisonersblog.pdf>.

36 In 2007 the Rudd Labor Government committed $500 million over seven years to CCS in the so-called Clean Energy Fund. In 2008, while in government, it committed an additional $100 million per year for an unspecified period for a

so-called Clean Coal Institute and in January 2009 another $1000 million for prototype coal-fired power stations with CCS. In 2007 Queensland committed $300 million to CCS. The New South Wales and Victorian governments have committed a total of $100–$200 million to CCS.

37 Ansolabehere, S et al (2007) *The Future of Coal: An interdisciplinary MIT study*, <http://web.mit.edu/coal>.

38 Sovacool, BK (2008) Valuing the greenhouse gas emissions from nuclear power: A critical survey, *Energy Policy* 36 (8): 2940–53; ISA (2006) *Life Cycle Energy Balance and Greenhouse Gas Emissions of Nuclear Energy in Australia*, <http://www.isa.org.usyd.edu.au>.

39 Van Leeuwen, JWS (2008) *Nuclear Power – The Energy Balance*, <http://www.stormsmith.nl>.

40 Hansen, J (2008) Tell Barack Obama the truth – the whole truth, 29 November, <http://www.columbia.edu/~jeh1/>.

41 Nuclear France <http://www.industcards.com/nuclear-france.htm>.

42 The 'fast' refers to the speed of the neutrons produced. For the distinction between a fast (neutron) reactor and a fast breeder reactor, see Glossary.

43 Ansolabehere et al (2003) *The Future of Nuclear Power: An interdisciplinary MIT study*, <http://web.mit.edu/nuclearpower>.

44 Diesendorf (2007a) chapter 12.

45 Digges, C (2007) UK government probe cites staff negligence in 2005 Thorp radioactive leak, 3 March, <http://www.bellona.org/articles/articles_2007/thorp_hse>.

46 For a Q & A by a nuclear physicist proponent, see Stanford, GS (2001) <http://www.nationalcenter.org/NPA378.html>.

47 Toke, D (2005) It strains the logic of energy economics, *The Guardian*, 5 October, <http://www.guardian.co.uk/society/2005/oct/05/guardiansocietysupplement8>.

48 Savage, M (2008) Power failure: What Britain should learn from Finland's nuclear saga, *Independent*, 16 January, <http://www.independent.co.uk/news/science/power-failure-what-britain-should-learn-from-finlands-nuclear-saga-770474.html>.

49 Edwards, G (1998) Findings on plutonium, Canadian Coalition for Nuclear Responsibility, <http://www.ccnr.org/Findings_plute.html>.

50 References in Diesendorf (2007a) chapter 12.

51 See Diesendorf (2007a) chapter 5.

52 Chang, AB, Rosenfeld, AH & McAuliffe, PK (2007) Energy efficiency in California and the United States: Reducing energy costs and greenhouse gas emissions, Figure 7, <http://www.energy.ca.gov/2007publications/CEC-999-2007-007>.

53 The global scenario is: McKinsey & Company (2009) *Pathways to a Low-Carbon Economy*, Version 2 of the global greenhouse gas *abatement* cost curve, <http://www.mckinsey.com/clientservice/ccsi/pathways_low_carbon_economy.asp>. The Australian scenario is McKinsey & Company (2008) *An Australian Cost Curve for Greenhouse Gas Reduction*, <http://www.greenfleet.com.au/uploads/pdfs/McKinsey%20Report%20-%20greenhouse%20-%2015Feb08.pdf>.

54 I have simplified the explanation by ignoring intermediate-load, which doesn't change the basic argument.

55 Because the quantity of water stored in a dam for hydroelectricity is limited, it too has an economic value, which is reflected in a 'fuel' cost.

56 I explain this in more detail in Diesendorf (2007a) chapter 4.

57 Additional power from combined cycle gas-fired power stations would also be required to substitute for the dawn to midnight contribution of those coal-fired power stations.

58 For more detail, see Diesendorf (2007a) and Diesendorf, M (2007c) *The Base Load Fallacy*, briefing paper no 16, <http://www.energyscience.org.au>.

59 Diesendorf, M (2007d) Submission to Garnaut Climate Change Review, Appendix B, <http://www.sustainabilitycentre.com.au/popular.html>.

60 Rabl, A & Spadaro, J (2000) Public health impact of air pollution and implications for the energy system, *Annual Reviews of Energy and the Environment* 25: 601–27. See also ExternE (Externalities of Energy, a European Commission study) <http://www.externe.info>.

61 Stern, N (2006) *Stern Review: The economics of climate change*, October, <http://www.occ.gov.uk/activities/stern.htm>.

62 McKinsey & Company (2008).

63 ABC (2009).

64 Stern (2006).

65 Australian Treasury (2008) *Australia's Low Pollution Future: The economics of climate change*, <http://www.treasury.gov.au/lowpollutionfuture>.

66 For example, see Hirst, E & Brown, M (1990) Closing the efficiency gap: Barriers to the efficient use of energy, *Resources, Conservation & Recycling* 3: 267–81; Grubb, MJ (1990) Energy efficiency and economic fallacies, *Energy Policy* 18: 783–85; Geller, H & Nadel, S (1994) Market transformation strategies to promote end-use efficiency, *Annual Review of Energy and the Environment* 19: 301–46; Sanstad, AH & Howard, RB (1994) 'Normal' markets, market imperfections and energy efficiency, *Energy Policy* 22: 811–18.

67 Some climate scientists and activists believe that global climate change has already reached the stage where emergency action is needed. I agree.

68 MacGill, I, Watt, M & Passey, R (2002) *The Economic Development Potential and Job Creation Potential of Renewable Energy: Australian case studies*, commissioned by Australian Cooperative Research Centre for Renewable Energy Policy Group, Australian Ecogeneration Association and Renewable Energy Generators Association; Diesendorf, M (2004) Comparison of employment potential of the coal and wind power industries, *International Journal of Environment, Workplace and Employment* 1: 82–90.

69 BMU (2006) *Ecological Industrial Policy: Memorandum for a 'new deal' for the economy, environment and employment*, Federal Ministry for the Environment, Nature Conservation and Nuclear Safety (BMU), Germany, pp 17–18, <http://www.erneuerbare-energien.de/inhalt/38345/20119>; Renner, M (nd) *Working for People and the Environment*, Worldwatch Report, <http://www.worldwatch.org/node/5925>; ACF & ACTU (2008) *Green Gold Rush: How ambitious environmental policy can make Australia a leader in the global race for green jobs*,

Australian Conservation Foundation & Australian Council of Trade Unions, Melbourne, pp 23–24, <http://www.acfonline.org.au/articles/news.asp?news_id=2047>.

70 An emissions trading scheme only fosters the economic optimum mix of technologies for a given carbon price for technologies that are in competitive markets. In general, efficient energy use and infrastructure (such as new transmission lines) required by renewable energy systems are not provided by competitive markets.

71 Betz, R & Sato, M (2006) Emissions trading: Lessons learnt from the 1st phase of the EU ETS [European Union Emission Trading Scheme] and prospects for the 2nd phase, Editorial, *Climate Policy* 6: 351–59; Ellerman, AD & Joskow, PL (2008) *The European Union's Emissions Trading Scheme in Perspective*, Pew Center on Global Climate Change, <http://www.pewclimate.org/publications>.

72 Betz & Sato (2006).

73 Australian Government (2008). Trading would start mid-2012.

74 The Australian Government policy allows an increase in the 2020 target to up to 25 per cent in the unlikely event that 'the world agrees to an ambitious global deal to stabilise levels of CO_2-equivalent in the atmosphere at 450 parts per million or less by 2050'. See 2009 fact sheet *New Measures for the CPRS* at <http://www.climatechange.gov.au/whitepaper/measures/index.html>.

75 Australian Government (2008).

76 Garnaut Review (2008) chapter 12. The 5 per cent reduction target recommended by Garnaut is not mentioned in the Key Points of chapter 12, perhaps because of the author's embarrassment at this very weak target, however it appears on p 282.

77 Global Commons Institute <http://www.gci.org.uk/contconv/cc.html>.

78 Hawken, P, Lovins, AB & Lovins, LH (2000) *Natural Capitalism: The next industrial revolution*, Earthscan, London.

79 Turton, H (2002) *The Aluminium Smelting Industry: Structure, market power, subsidies and greenhouse gas emissions*, discussion paper 44, Australia Institute, Canberra, Section 2.7.

80 Even after 50 years, nuclear power can be described as immature, because it is still not 'fail-safe', that is, designed to be completely safe in the event of an accident.

Chapter 3: Technologies for stopping global warming

1 Diesendorf, M (2007a) *Greenhouse Solutions with Sustainable Energy*, UNSW Press, Sydney.

2 Boyle, G (ed) (2004) *Renewable Energy: Power for a sustainable future*, 2nd ed, Open University and Oxford University Press, Oxford. (This is a well-illustrated introductory textbook for lay readers and scientists alike.)

3 Sørensen, B (2005) *Renewable Energy: Its physics, engineering, environmental impacts, economics and planning*, 3rd ed, Academic Press, San Diego. (This is a comprehensive in-depth textbook for science and engineering students.)

4 This is a project of the Centre for Sustainable Energy Systems at the Australian National University, see <http://solar.anu.edu.au/projects/chaps_proj.php>.

5   Global Wind Energy Council <http://www.gwec.net>.
6   NREL (2008) *20% Wind Energy by 2030: Increasing wind energy's contribution to US electricity supply*, National Renewable Energy Laboratory for US Department of Energy, May, <http://www.osti.gov/bridge>.
7   This and other fallacies about wind power are refuted in detail in Diesendorf (2007a) chapter 6.
8   Diesendorf (2007a) chapter 6.
9   Diesendorf (2007a) chapter 7.
10 However, the production of ethanol from corn in the USA and the production of biodiesel from palm oil obtained by clearing tropical rainforest is not carbon neutral (see 'Transport solutions' section in this chapter).
11 Saddler, H, Diesendorf, M & Denniss, R (2007) Clean energy scenarios for Australia, *Energy Policy* 35(2): 1245–56.
12 Research Institute for Sustainable Energy (RISE), Concentrated solar, <http://www.rise.org.au/info/Tech/hightemp/index.html>; California, Office of the Governor (2008) Media release, 23 October, <http://gov.ca.gov/press-release/10876>.
13 RISE, Photovoltaics <http://www.rise.org.au/info/Tech/pv/index.html>.
14 CSG Solar <http://www.csgsolar.com>.
15 Sliver Technology Research at the ANU <http://solar.anu.edu.au/research/sliver.php>.
16 Géothermie Soultz <http://www.soultz.net/fr>.
17 Geodynamics Limited <http://www.geodynamics.com.au>.
18 MIT (2006) *The Future of Geothermal Energy*, Massachusetts Institute of Technology, <http://www1.eere.energy.gov/geothermal/future_geothermal.html>.
19 International Rivers <http://internationalrivers.org>. This global network protects rivers and the communities that depend upon them.
20 RISE, Tidal barrage and tidal turbines, <http://www.rise.org.au/info/Tech/tidal/index.html>; RISE, Wave, <http://www.rise.org.au/info/Tech/wave/index.html>.
21 Rosenthal, E (2008) New trend in biofuels has new risks, *New York Times*, 21 May, <http://www.nytimes.com/2008/05/21/science/earth/21biofuels.html?_r=2>.
22 Victoria Transport Policy Institute <http://www.vtpi.org/tdm/index.php>.
23 Newman, P & Kenworthy, J (2006) Urban design to reduce automobile dependence, *Opolis: An International Journal of Suburban and Metropolitan Studies* 2 (1): Article 3, <http://repositories.cdlib.org/cssd/opolis/vol2/iss1/art3>. The conceptual plan is summarised and illustrated in Diesendorf (2007a) chapter 10. See also Newman, P & Kenworthy, J (1999) *Sustainability and Cities: Overcoming automobile dependence*, Island Press, Washington DC. An earlier conception of transforming cities into clusters of clusters is proposed in White, D, Sutton, P, Pears, A, Dick, J & Crow, M (1978) *Seeds for Change: Creatively confronting the energy crisis*, Patchwork Press and Conservation Council of Victoria, Melbourne.
24 Patel-Predd, P (2008) A battery-capacitor hybrid – for hybrids, *ieee spectrum online*, December, <http://www.spectrum.ieee.org/dec08/6991>.

25 Intergovernmental Panel on Climate Change (IPCC) (2007d) *Climate Change 2007: Mitigation of climate change*, Technical Summary, chapter 8, <http://www.ipcc.ch>.

26 Saddler, H & King, H (2008) *Agriculture and Emissions Trading: The impossible dream*, Australia Institute, discussion paper 102, <http://www.tai.org.au>.

27 Marris, E (2006) Black is the new green, *Nature* 442: 624–26.

28 Parr, JF & Sullivan, LA (2005) Soil carbon sequestration in phytoliths, *Soil Biology & Biochemistry* 37: 117–24.

29 Sørensen, B with contributions from Kuemmel, B & Meibom, P (1999) *Long-Term Scenarios for Global Energy Demand and Supply: Four global greenhouse mitigation scenarios*, Final Report, Tekst Nr 359, IMFUFA, Roskilde Universitetscenter. A summary is available in Sørensen (2005).

30 Sørensen, B & Meibom, P (2000) A global renewable energy scenario, *International Journal of Global Energy Issues* 13(1/2/3), DOI: 10.1504/IJGEI.2000.000869; a summary is available in Sørensen (2005).

31 McKinsey & Company (2009) *Pathways to a Low-Carbon Economy*, Version 2 of the global greenhouse gas abatement cost curve, <http://www.mckinsey.com/clientservice/ccsi/pathways_low_carbon_economy.asp>.

32 These numbers are likely to be inaccurate because McKinsey does not tabulate them and so with difficulty I have attempted to read them from McKinsey's cost curve.

33 Makhijani, A (2007) *Carbon Free and Nuclear Free: A roadmap for US energy policy*, RDR Books, Muskegon MI and IEER Press, Takoma Park MD.

34 Jacobson, MZ (2009) Review of solutions to global warming, air pollution and energy security, *Energy & Environmental Science* 6(2): 148–93, DOI: 10.1039/b809990c.

35 Diesendorf, M (2007b) Paths to a Low-Carbon Future: Reducing Australia's greenhouse gas emissions by 30 per cent by 2020, Greenpeace Australia Pacific, September; Teske, S & Vincent, J (2008) Energy [R]evolution: A sustainable Australia energy outlook, Greenpeace Australia Pacific and European Renewable Energy Council, <http://www.energyblueprint.info>; McKinsey & Company (2008) An Australian Cost Curve for Greenhouse Gas Reduction, <http://www.greenfleet.com.au/uploads/pdfs/McKinsey%20Report%20-%20greenhouse%20-%2015Feb08.pdf>.

36 Martinot, E & Junfeng, L (2007) *Powering China's Development: The role of renewable energy*, Worldwatch Institute <http://www.worldwatch.org/node/5491>; Climate Group (2008) *China's Clean Revolution*, <http://www.theclimategroup.org/assets/resources/Chinas_Clean_Revolution.pdf>; *China Renewable Energy and Sustainable Development Report* (2008) <http://www.energy-base.org/fileadmin/media/sefi/docs/industry_reports/China_Strategies__LLC--January_2008_China_Renewable_Energy_Report_in_Word_Format.pdf>; Sawin, JL & Martinot, E (2009) *Renewables Global Status Report: 2008 update*, Worldwatch Institute <http://www.worldwatch.org>.

37 Brown, L (2008) *Plan B 3.0: Mobilizing to save civilization*, 3rd ed, WW Norton, New York.

38 Spratt, D and Sutton, P (2008) *Climate Code Red: The case for emergency action*,

Scribe, Melbourne.

39 Although the loser may suffer major destruction of its economy, that doesn't invalidate the argument.

Chapter 4: Essential policies for the 21st century

1   Beder, S (2000) *Global Spin: The corporate assault on environmentalism*, revised ed, Scribe, Melbourne.

2   Intergovernmental Panel on Climate Change (IPCC) (2007c) *Climate Change 2007: Synthesis report*, An assessment of the Intergovernmental Panel on Climate Change, table 5.1.

3   Hansen, J, Sato, M, Kharecha, P et al (2008) Target atmospheric CO_2: Where should humanity aim? *Open Atmospheric Science Journal* 2: 217–31.

4   See Diesendorf (2007b) *Paths to a Low-Carbon Future: Reducing Australia's greenhouse gas emissions by 30 per cent by 2020*, Greenpeace Australia Pacific, September; and McKinsey & Company (2008) *An Australian Cost Curve for Greenhouse Gas Reduction*, <http://www.greenfleet.com.au/uploads/pdfs/McKinsey%20Report%20-%20greenhouse%20-%2015Feb08.pdf>.

5   Teske, S & Vincent, J (2008) *Energy [R]evolution: A sustainable Australia energy outlook*, Greenpeace Australia Pacific and European Renewable Energy Council, <http://www.energyblueprint.info>.

6   Australian Government (2008) *Carbon Pollution Reduction Scheme: Australia's low pollution future*, White Paper, Canberra, table E1; European Parliament (2008) *European parliament seals climate change package*, Press release, 17 December, <http://www.europarl.europa.eu/news/expert/infopress_page/064-44858-350-12-51-911-20081216IPR44857-15-12-2008-2008-false/default_en.htm>; Obama, B & Biden, J (nd) *New Energy for America*, Fact sheet <http://my.barackobama.com/page/content/newenergy>; AAP (2008) Mexico pledges 50 per cent cut in greenhouse gases, *International Herald Tribune*, 11 December, <http://www.iht.com/articles/ap/2008/12/11/europe/EU-Poland-Climate-Mexico.php>.

7   Strictly speaking, CO_2 is emitted in manufacturing renewable energy technologies, however this is similar in magnitude to the emissions from manufacturing fossil fuel technologies that generate the same quantity of energy and considerably less than the emissions from the nuclear fuel chain. As energy supply becomes more and more renewable, CO_2 emissions from manufacturing renewable energy technologies will become negligible.

8   Diesendorf (2007b); Saddler, H, Diesendorf, M & Denniss, R (2004) *A Clean Energy Future for Australia*, Clean Energy Future Group, Sydney; Saddler, H, Diesendorf, M & Denniss, R (2007) Clean energy scenarios for Australia, *Energy Policy* 35(2): 1245–56.

9   This target is less impressive than it looks, because at least half can be met by offsets in non-EU countries. In 2008, renewable energy was 8.5 per cent of the EU's primary energy.

10 The UK is also bound by EU targets, however its share is unclear. In 2007, renewable energy provided 1.8 per cent of UK primary energy and 4 per cent of electricity.

11 Obama & Biden (nd) pre-election promise.
12 In 2008, renewable energy provided about 8.5 per cent of Australia's electricity generation. The vast majority is hydro.
13 Repower America <http://www.repoweramerica.org>. Repower America's concept of 'clean energy' includes nuclear power contributing 24 per cent of electricity by 2020.
14 Hansen, Sato, Kharecha et al (2008) figure 6 and associated text.
15 For example, they could be the integrated gasification combined cycle type of power station, which is just coming onto the market. However, without CCS, its emissions will be similar to existing power stations that burn pulverised coal.
16 New Rules Project, Democratic energy, <http://www.newrules.org/environment/climateca2.html>.
17 Regional Greenhouse Gas Initiative <http://www.rggi.org/home>.
18 Obama & Biden (nd).
19 Australian Government (2008) chapter 13.
20 Fryer, D, Barraclough, M & Crooks, R (2008) *The Impact of Industries Assistance Measures under the Carbon Pollution Reduction Scheme – White Paper Update*, Innovest, in Australian Conservation Foundation, Households to foot the big polluters' carbon bill, <http://www.acfonline.org.au/articles/news.asp?news_id=2103>.
21 Saddler, H, Muller, F & Cuevas, C (2006) *Competitiveness and Carbon Pricing: Border adjustments for greenhouse policies*, discussion paper 86, Australia Institute, Canberra, April.
22 Extracted from MacGill, I & Betz, R (2008) *Emissions trading: Good governance requires 100% auctioning*, Centre for Policy Development <http://cpd.org.au/article/emissions-trading-good-governance-requires-100-auctioning>.
23 Point Carbon (2008) *Carbon 2008: Post–2012 is now*, <http://www.pointcarbon.com/research/carbonmarketresearch/analyst/1.912721>. In phase I the excess allocation was about 100 million tones of CO_2; in phase II the allocation was about 13 per cent below the 2005 allocation.
24 Rosenthal, E (2008) Europe turns back to coal, raising climate fears, *New York Times*, 23 April, <http://www.nytimes.com/2008/04/23/world/europe/23coal.html>; Paterson, T (2007) German Greens fight coal-fired power station plan, *The Independent*, 23 March, <http://www.independent.co.uk/news/world/europe/german-greens-fight-coalfired-power-station-plan-441477.html>; Reuters UK (2007) Table–Coal-fired power station projects in Europe, 14 May, <http://uk.reuters.com/article/oilRpt/idUKL1073510020070514>.
25 Surprisingly, the Australian ETS does not cover deforestation, a significant source of emissions.
26 Saddler, H & King, H (2008) *Agriculture and Emissions Trading: The impossible dream*, discussion paper 102, Australia Institute, Canberra, <http://www.tai.org.au>.
27 UNFCC (2007) The Kyoto Protocol Mechanisms: International Emissions Trading, Clean Development Mechanism, Joint Implementation, <http://unfccc.int/resource/docs/publications/mechanisms.pdf>.
28 Haya, B (2007) *Failed Mechanism: How the CDM is subsidizing hydro developers*

and harming the Kyoto Protocol, International Rivers, Berkeley CA, <http://internationalrivers.org/files/Failed_Mechanism_3.pdf>.

29 Schleich, J, Rogge, K & Betz, R (2008) *Incentives for Energy Efficiency in the EU Emissions Trading Scheme,* Working Paper Sustainability and Innovation, No S 2/2008, Fraunhofer Institute for Systems and Innovation Research.

30 De Moor, A (2001) Towards a grand deal on subsidies and climate change, *Natural Resources Forum* 25(2): 167–76.

31 This was expressed in US 1990 dollars. Anderson, K (1995) The political economy of coal subsidies in Europe, *Energy Policy* 23: 485–96.

32 Roder, A (2003) *An Overview of Senate Energy Bill Subsidies to the Fossil fuel Industry,* Taxpayers for Commonsense, <http://archives.eesi.org/briefings/2003/Special%20Issues/5.12.03%20Fossil%20Subsidies/5.12.03Roder%20Comments.doc>.

33 Riedy, C & Diesendorf, M (2003) Financial subsidies and incentives to the Australian fossil fuel industry, *Energy Policy* 31 (2): 125–37; Riedy, CJ (2007) *Energy and Transport Subsidies in Australia: 2007 update,* Institute for Sustainable Futures, Sydney.

34 Buckman, G & Diesendorf, M (2009) *The future of renewable electricity in Australia,* submitted for publication.

35 Mandatory Renewable Energy Target Review <http://www.mretreview.gov.au>.

36 From 2000 to 2010, the target was increased in steps up to 9500 GWh per year.

37 Australian Department of Climate Change (2008) *Design of the renewable energy target (RET) scheme – release of exposure draft legislation,* Fact sheet, <http://www.climatechange.gov.au/renewabletarget/publications/pubs/fs-ret-exposure-draft-legislation.pdf>.

38 Pacific Hydro (2009) *Submission to the Renewable Energy Target Exposure Draft Legislation,* <http://www.pacifichydro.com.au/Portals/0/RET%20Exposure%20Legislation%20Submission%20PHL.pdf>.

39 South Australia, Victoria and Queensland are implementing net FITs. The Australian Capital Territory has a gross FIT and Western Australia is considering one.

40 European Wind Energy Association <http://www.ewea.org>.

41 This is done in the Australian Capital Territory.

42 Brotherhood of St Laurence & KPMG (2008) *A National Energy Efficiency Program to assist Low-Income Households,* September, <http://www.bsl.org.au/main.asp?PageId=5394>.

43 US Department of Energy, Smart Grid, <http://www.oe.energy.gov/smartgrid.htm>.

44 Mackey, B, Keith, H, Berry, S & Lindenmayer, DB (2008) *Green Carbon: The role of natural forests in carbon storage,* Part 1. A green carbon account of the eucalypt forests of south-east Australia. ANU E Press, Canberra.

45 Blakers, M (2008) *Biocarbon, Biodiversity and Climate Change: A REDD plus scheme for Australia,* Green Institute Working Paper 3, <http://www.greeninstitute.com.au/content/index.php?/site/projects/forests_and_greenhouse>.

46 Forest Stewardship Council <http://www.fsc.org>; FSC Watch <http://www.fsc-

watch.org>.

47 Saddler & King (2008).

48 Glover, JD, Cox, CM & Reganold, JP (2007) Future farming a return to roots?, *Scientific American* 297(2): 82–89, <http://www.landinstitute.org/pages/Glover-et-al-2007-Sci-Am.pdf>.

49 Briggs, C, Cole, M, Evesson, J et al (2007) *Going with the Grain?: Skills and sustainable business development*, NSW Board of Vocational Education and Training, <http://www.bvet.nsw.gov.au/pdf/GoingWithTheGrain.pdf>; Hatfield-Dodds, S, Turner, G, Schandl, H & Doss, T (2008) *Growing the Green Collar Economy: Skills and labour challenges in reducing our greenhouse emissions and national environmental footprint*, Report to the Dusseldorp Skills Forum, CSIRO Sustainable Ecosystems, Canberra, June.

50 'Have another child for the country' stated by then Australian treasurer Peter Costello.

51 Bloom, DE & Canning, D (2008) Global demographic change: dimensions and economic significance, in Preskawetz, A, Bloom, DE & Lutz, W (eds) *Population Aging, Human Capital Accumulation, and Productivity Growth*, Population and Development Review, Supplement to vol 34, p 20, Population Council, New York.

52 O'Connor, M & Lines, WJ (2008) *Overloading Australia: How governments and media dither and deny on population*, Envirobook, Sydney, chapter 20.

53 Blatt, JM (1983) *Dynamic Economic Systems*, ME Sharpe, Armonk NY, and Wheatsheaf, Brighton, Sussex; Diesendorf, M & Hamilton, C (eds) *Human Ecology Human Economy: Ideas for an ecologically sustainable future*, Allen & Unwin, Sydney, especially chapter 2 by Clive Hamilton on 'Foundations of ecological economics'; Ormerod, P (1998) *Butterfly Economics*, Faber & Faber, London; Keen, S (2001) *Debunking Economics: The naked emperor of the social sciences*, Pluto Press, Sydney; Davies, G (2004) *Economia: New economic systems to empower people and support the living world*, ABC Books, Sydney; Green, D (2008) *From Poverty to Power: How active citizens and effective states can change the world*, Oxfam International; New Economics Foundation <http://www.neweconomics.org/gen>.

54 See list of books in previous note.

55 In neoclassical economics, this assumption is called 'decreasing or constant returns to scale'.

56 Davies (2004).

57 For an introduction, see Wikipedia (2009) Fractional Reserve Banking, 2 April, <http://en.wikipedia.org/wiki/Fractional-reserve_banking>.

58 Daly, H & Cobb, J (1989) *For the Common Good: Redirecting the economy toward community, the environment and a sustainable future*, Beacon Press, Boston; Hamilton, C (1999) The genuine progress indicator: Methodological developments and results from Australia, *Ecological Economics* 30: 13–28.

59 Davies (2004) part 7.

60 Global Reporting Initiative <http://www.globalreporting.org>.

61 Berger, C (2006) *False Profits: How Australia's finance sector undervalues the environment and what we can do about it*, Australian Conservation Foundation,

Melbourne <http://www.acfonline.org.au/uploads/res/res_false_profits.pdf>.
62 ICLEI – Local Governments for Sustainability <http://www.iclei.org>.

Chapter 5: Strategies for defeating the Greenhouse Mafia

1   Sharp, G (nd) *The Importance of Strategic Planning in Nonviolent Struggle*,
    <http://www.nonviolence.org.au/downloads/strategic_planning.pdf>.
2   Bobo, KA, Kendall, J & Max, S (2001) *Organising for Social Change: Midwest Academy manual for activists*, 3rd ed, Seven Locks Press, Washington.
3   Climate Institute (2007) *Climate Institute: Marginal electorates exit poll*, conducted by Australian Research Group Pty Ltd, November, <http://www.climateinstitute.org.au/images/stories/exitpoll.pdf>.
4   Climate Institute (2008) *Government's climate change credentials slump*, 23 October, conducted by Auspoll, search <http://www.climateinstitute.org.au>.
5   Alinsky, SD (1971) *Rules for Radicals*, Random House, New York, p 113.
6   Sharp, G (1973) *The Politics of Nonviolent Action*, Porter Sargent, Boston, p 3.
7   The Right Livelihood Award <http://www.rightlivelihood.org>; Ekins, P (1992) *A New World Order: Grassroots movements for global change*, Routledge, London.
8   Gibbs, LM (1998) *Love Canal: The story continues*, New Society Publishers, Stony Creek CT, USA.
9   This list is modified slightly from that of Bobo et al (2001).
10 Rose, C (2005) *How to Win Campaigns: 100 steps to success*, Earthscan, London.
11 The Wikipedia entry for Brent Spar (2009) <http://en.wikipedia.org/wiki/Brent_Spar> 10 January provides a good introduction.
12 Citizens' groups in Minamata played an important role in a long struggle that eventually identified the cause of their affliction as methyl mercury discharged into the local bay by the Chisso Corporation: see Ui, J (1992) *Industrial Pollution in Japan*, United Nations University Press, Tokyo, chapter 4 'Minamata disease'.
13 Engel, S and Martin, B (2006) Union Carbide and James Hardie: Lessons in politics and power, *Global Society* 20(4): 475–90.
14 Kneale, P & Aspinall, S (2006) *SWOT analysis – An Introduction*, <http://www.uk-student.net/modules/wfsection/article.php?articleid=91>; Houben, G, Lenie, K & Vanhoof, K (1999) A knowledge-based SWOT-analysis system as an instrument for strategic planning in small and medium-sized enterprises, *Decision Support Systems* 26(2): 125–35.
15 Bobo et al (2001) chapter 4.
16 Doyle, T (2000) *Green Power: The environment movement in Australia*, UNSW Press, Sydney.
17 Ceres <http://www.ceres.org>.
18 Australian Business Roundtable on Climate Change <http://www.businessroundtable.com.au>.
19 Central Victorian Greenhouse Alliance <http://www.cvga.org.au/main>.
20 Climate Institute (2008) *New powerful climate alliance to drive tough and fair solutions*, Media release, 7 July, search <http://www.climateinstitute.org.au>.
21 WWF-Australia, WWF welcomes National Low Emission Coal Council,

<http://wwf.org.au/news/wwf-welcomes-national-low-emission-coal-council>;
WWF joins world's leading environment proponents in CCS call, <http://wwf.
org.au/news/wwf-joins-worlds-leading-environment-proponents-in-ccs-call>.

22 United States Climate Action Partnership (USCAP) <http://www.us-cap.org/
index.asp>.

23 USCAP 'believe the construction of new plants should occur in a manner that
will allow them to capture and store CO_2 when the conditions exist to support
its implementation'. However, in theory, all coal-fired power stations could
have CCS added on for a price, so USCAP's position is not reassuring from an
environmental viewpoint.

24 See appendix 2 and Apollo Alliance <http://apolloalliance.org>.

25 Maddison, S & Scalmer, S (2006) *Activist Wisdom: Practical knowledge
and creative tension in the life of social movements*, UNSW Press, Sydney,
pp 151–52.

26 Hall & Taplin (2007a).

27 National Leadership Summits for a Sustainable America, *Wingspread Principles
on the US Response to Global Warming*, <http://www.summits.ncat.org/energy_
climate/statement.php>. Signatories include: Ray Anderson, Chairman of the
Board, Interface Inc; Roger Duncan, Deputy General Manager, Austin Energy,
Austin, Texas; Jonathan Lash, President, World Resources Institute; L Hunter
Lovins, President, Natural Capitalism Solutions; Michael C MacCracken, PhD,
Chief Scientist for Climate Change Programs, Climate Institute, Washington
DC; Paul Raskin, PhD, President, Tellus Institute; Rob Sargent, Energy
Program Director, Environment America; Larry J Schweiger, President and
CEO, National Wildlife Federation; Randall Swisher, Executive Director,
American Wind Energy Association; Mike Tidwell, Director, Chesapeake
Climate Action Network; US Green Building Council, Board of Directors;
John R Wells, Strategic Planning Director, Minnesota Environmental Quality
Board; CL Winter, PhD, Deputy Director, National Center for Atmospheric
Research, Boulder, CO; Michelle Wyman, Executive Director, ICLEI – Local
Governments for Sustainability, USA, Inc; plus numerous professors and
mayors.

28 Fuller, AA (nd) What works? Evidence from research on nonviolent social
movements, *Bulletin of the Manchester College Peace Institute*, <http://aer.
manchester.edu/Academics/Departments/Peace_Studies/bulletin/2008/
documents/samplePDF.pdf>

29 Gamson, WA (1975, 1990) *The Strategy of Social Protest*, Dorsey Press &
Wadsworth, Homewood, Ill.

30 Cress, D & Snow, DA (1992) Mobilization at the margins: resources,
benefactors and the viability of homeless social movement organizations,
American Sociological Review 61(6): 1089–109.

31 Martin, B (1991) Social defence: arguments and actions, in Anderson, S
& Larmore, J (eds), *Nonviolent Struggle and Social Defence*, War Resisters'
International and the Myrtle Solomon Memorial Fund Subcommittee, London,
pp 81–83, <http://www.uow.edu.au/arts/sts/bmartin/pubs/91nssd/intro.html>.

32 One reader of this chapter, who is both an academic and an activist, commented

that 'you are unnecessarily being too prescriptive about the need for organisation and preferred structures', referring me to Jordan, T (2004) *Activism: Direct action, hacktivism and the future of society*, University of Chicago Press, Chicago.

33 Fuller (nd).

34 See Alinsky (1971) and his earlier book: Alinsky, SD (1946, 1969) *Reveille for Radicals*, Vintage Books, New York.

35 Hall, N, Taplin, R & Goldstein, W (2009, in press) Empowerment of individuals and realisation of community agency: Applying Action Research to climate change responses in Australia, *Action Research*.

36 Hall, Taplin & Goldstein (2009).

37 Moyer, B (1987) *The Movement Action Plan*, <http://www.indybay.org/olduploads/movement_action_plan.pdf>; Moyer, B, with McAllister, J, Finley, ML & Soifer, S (2001) *Doing Democracy: The MAP model for organizing social movements*, New Society Publishers, Gabriola Island BC, Canada.

38 Moyer et al (2001).

39 Moyer (1987); Moyer et al (2001).

40 Climate Institute (2007, 2008).

41 Zogby International (2006) <http://www.zogby.com/wildlife/NWFfinalreport 8-17-06.htm>.

42 Pew Research Center for People and the Press (2006) *Little consensus on global warming*, <http://people-press.org/report/280/little-consensus-on-global-warming>.

43 Diesendorf, M (2007a) *Greenhouse Solutions with Sustainable Energy*, UNSW Press, Sydney, pp 330–32.

44 Garnaut, R (2008) *The Garnaut Climate Change Review: Final report*, Cambridge University Press, Cambridge, <http://www.garnautreview.org.au/pdf/Garnaut_prelims.pdf>.

45 Hall, NL & Taplin, R (2008) Room for climate advocates in a coal-focused economy? NGO influence on Australia climate policy, *Australian Journal of Social Issues* 43: 359–79.

46 Regional Greenhouse Gas Initiative <http://www.rggi.org/home>.

47 Pew Center on Global Climate Change <http://www.pewclimate.org/about>.

48 The Climate Institute describes itself as a 'think-tank'. However, it also fits our broad definition of a CMO.

49 ACTU <http://www.actu.asn.au>.

50 ACF & ACTU (2008) *Green Gold Rush: How ambitious environmental policy can make Australia a leader in the global race for green jobs*, Australian Conservation Foundation and Australian Council of Trade Unions, October, <http://www.acfonline.org.au/uploads/res/Green_Gold_Rush_final.pdf>.

51 Climate Action Network <http://www.climatenetwork.org>.

52 Stop Climate Chaos <http://www.stopclimatechaos.org/about>.

53 Campaign Against Climate Change <http://www.campaigncc.org>.

54 Climate Action Network Australia <http://www.cana.net.au>.

55 Climatemovement.org.au <http://www.climatemovement.org.au>.

56 Australian Youth Climate Coalition <http://www.aycc.org.au>.

57 Australian Student Environment Network <http://asen.org.au>.

58 Beder, S (2000) *Global Spin: The corporate assault on environmentalism*, revised ed, Scribe Publications, Melbourne, especially chapter 14.

59 Australian Workers Union <http://www.awu.net.au>.

60 CFMEU (2006) *CFMEU targets mining companies in campaign to reduce global warming*, Media release, 16 November, <http://www.cfmeu.com.au/index.cfm?section=5&Category=42&viewmode=content&contentid=25>.

61 Maitland, J (2003) *Challenges Facing the Coal Industry*, Address to CEDA Industry Update, July, <http://www.cfmeu.com.au/storage/documents/JM_CEDA2003.pdf>.

62 CFMEU (2005) *Response to NSW Energy Directions Green Paper*, Construction, Forestry, Mining and Energy Union, February, <http://www.cfmeu.com.au/storage/documents/energysub05.pdf>.

Chapter 6: Winning campaign tactics

1 Since I'm not familiar with all of these groups, this list is not necessarily a recommendation.

2 Environmental Defender's Office (NSW) Ltd (2007) *Campaigning and the Law in New South Wales: A guide to your rights and responsibilities*, <http://www.edo.org.au/edonsw/site/pdf/pubs/campaigning.pdf>. The EDO is an Australian community legal centre specialising in environmental law.

3 Bellamy, D (2005) Glaciers are cool, *New Scientist* 2495, 16 April; refuted by Monbiot, G (2005) Junk science, *The Guardian*, 10 May, <http://www.monbiot.com/archives/2005/05/10/junk-science>.

4 Rose, C (2005) *How To Win Campaigns*, Earthscan, London, chapter 6.

5 Riedy, C & Diesendorf, M (2003) Financial subsidies to the Australian fossil fuel industry, *Energy Policy* 31: 125–37; updated in Riedy, CJ (2007) *Energy and Transport Subsidies in Australia: 2007 update*, Institute for Sustainable Futures, Sydney.

6 George Lakoff interviewed by Katy Butler for the Sierra Club <http://www.sierraclub.org/sierra/200407/words.asp>. For a thorough discussion of 'framing', see Lakoff, G (2004) *Don't Think Like an Elephant!: Know your values and frame the debate*, Chelsea Green, White River Junction VT.

7 Beder, S (2000) *Global Spin: The corporate assault on environmentalism*, revised ed, Scribe Publications, Melbourne.

8 Australian Government (2006) *Uranium Mining, Processing and Nuclear Energy Review*, archived at <http://pandora.nla.gov.au/tep/66043>.

9 RMIT Community Advocacy Unit (2008) *Lobbying Know How Resource Kit*, RMIT, Melbourne.

10 In south-eastern New South Wales, Clean Energy for Eternity initiated a project to power a local surf clubhouse with renewable energy. Subsequently this turned into a national program, transforming a national icon, surf lifesaving, into a 'green' activity.

11 OOA <http://www.ooa.dk/eng/engelsk.htm>.

12 Australian Science Media Centre (2008) *Rapid Roundup: Carbon Pollution Reduction Scheme – White Paper – experts respond*, <http://www.aussmc.org/CPRS_White_Paper.php>.

13 Clean Energy for Eternity <http://austcom.org.au/cefe.html>.
14 Marshall, C (2009) Death by sound bites? The language of the cap-and-trade debate, *New York Times*, 9 March, <http://www.nytimes.com/cwire/2009/03/09/09climatewire-death-by-sound-bites-the-language-of-the-capa-9991.html>.
15 Rose (2005) chapter 7.
16 Rose (2005).
17 King, ML Jr (1964) *Why Can't We Wait*, Signet, New York, p 12.
18 Gandhi, MK (1927, 1929) *An Autobiography or The Story of my Experiments with Truth*, Navajivan Publishing House, Ahmedabad; Merton, T (1964) *Gandhi on Non-Violence*, New Directions, New York.
19 There are many documented cases, eg, Churchill, W & Vander Wall, J (1988, 2002) *Agents of Repression: The FBI's secret wars against the black panther party and the American Indian movement*, South End Press, Boston; Ganser, D (2004) *NATO's Secret Armies: Operation Gladio and terrorism in Western Europe*, Routledge, New York.
20 Coover, V, Deacon, E, Esser, C & Moore, C (1978) *Resource Manual for a Living Revolution*, 2nd ed, Movement for a New Society, Philadelphia.
21 These actions could be unlawful if unauthorised. Relevant offences may include those relating to obstruction, unlawful assembly, intimidation, engaging in dangerous activities, and interference with rights. Civil laws that may apply include those relating to nuisance, trespass, interference with use of land, interference with contract, interference with business, intimidation, invasion of privacy and conspiracy. Check the law in your jurisdiction before proceeding.
22 Adam, D (2006) Royal Society tells Exxon: Stop funding climate change denial, *The Guardian*, 20 September, <http://environment.guardian.co.uk/print/0,,329580967-121568,00.html>.
23 Other civil laws may also apply, such as injurious falsehood. Furthermore, the recent 'Gunns 20' litigation in Australia <http://www.gunns20.org> is an example of how a corporation might try to sue campaigners for statements made against it. These kinds of cases are often referred to as 'SLAPP suits' (Strategic Litigation Against Public Participation).
24 That was the case for a previous Australian Minister for Energy.
25 Alinsky, SD (1971) *Rules for Radicals*, Random House, New York, pp 170–83.
26 Strikes and boycotts can have legal implications, including breaches of industrial laws or laws regulating trade practices, although some boycotts may be lawful where the dominant purpose is environmental protection. Businesses might also try to sue campaigners for economic loss suffered, through what are known as 'economic torts'.
27 Sharp, G (1973) *The Politics of Nonviolent Action*, Porter Sargent, Boston.
28 For background information and to see how the case unfolded visit <http://www.sierraclub.org/coal/plantlist.asp>. A copy of the decision can be found at: <http://yosemite.epa.gov/oa/EAB_Web_Docket.nsf/PSD%20Permit%20Appeals%20(CAA)/C8C5985967D8096E85257500006811A7/$File/Remand...39.pdf>.
29 McCarthy, M (2008) Cleared: Jury decides that threat of global warming

justifies breaking the law, *The Independent*, 11 September, <http://www.independent.co.uk/environment/climate-change/cleared-jury-decides-that-threat-of-global-warming-justifies-breaking-the-law-925561.html>.

30 Greenpeace UK (2008) *Government moves to strip juries of power in climate protest cases*, Media release, 18 December, <http://www.greenpeace.org.uk/media/press-releases/government-moves-strip-juries-its-power-climate-protest-cases-20081218>.

31 Hancock, WK (1974) *The Battle of Black Mountain: An episode of Canberra's environmental history*, Department of Economic History, Research School of Social Sciences, Australian National University, Canberra.

32 Westmill Co-op <http://www.westmill.coop/westmill_home.asp>.

33 Hepburn Renewable Energy Association <http://www.hrea.org.au>.

34 See appendix 1 and Clean Energy for Eternity <http://austcom.org.au/780.html>.

35 Clean Energy Council (2008) *Australia heads towards a clean energy future one budget at a time*, Media release, 13 May, <http://www.cleanenergycouncil.org.au/news/showarticle.php?id=120>.

36 Fyfe, M (2008) 'Dirty' fuel firms split clean energy group, *The Age*, 13 July, <http://www.theage.com.au/national/dirty-fuel-firms-split-clean-energy-group-20080712-3e6t.html>; Galacho, O (2008) Renewable energy body in turmoil, *Herald Sun*, 12 July.

37 BCA (2008) *Modelling Success: Designing an ETS that works*, Business Council of Australia, August, <http://www.bca.com.au/Content/101484.aspx>; ESAA (2006) *Energy and Emissions Study – Stage 2*, Energy Supply Association of Australia, November, <http://www.esaa.com.au/images/stories//energyandemissionsstudystage2.pdf>.

Appendix 1: Riding on trust and ripe conditions in the UK

1 Viral marketing uses pre-existing social networks to increase brand awareness, product sales (and other marketing objectives). It does this through self-replicating viral processes, analogous to the spread of pathological and computer viruses. It can be delivered via word-of-mouth or enhanced by the network effects of the internet. Barack Obama used viral campaigning to great effect during his push for presidency, see Tumulty, K (2007) Obama's viral marketing campaign, *Time*, 5 July, <http://www.time.com/time/magazine/article/0,9171,1640402,00.html>.

2 Editor (2008) This foolish rush into the arms of the dirtiest fuel, *The Independent*, 10 March, <http://www.independent.co.uk/opinion/leading-articles/leading-article-this-foolish-rush-into-the-arms-of-the-dirtiest-fuel-793661.html>.

GLOSSARY

The terms defined here are shown in italics where they first appear in the main text and in the definitions below. The definitions are within the context of climate action.

abatement See *mitigation*.

adaptation The technical term for reducing the impacts of climate change rather than the causes. Eg, building sea-walls; shifting farms from drought areas to those with higher rainfall; spraying artificial snow onto ski slopes; improving medical and public health facilities for treating the spread of malaria and dengue.

affluence Consumption (household expenditure) per person.

albedo The fraction of sunlight reflected by the Earth.

anthropogenic Caused by humans.

astroturfing The formation of front organisations by powerful vested interests.

automobile city A city in which most passenger transport is by automobile.

base-load The minimum daily level of power. Also describes power stations that operate at *rated power* 24 hours per day, 7 days a week, wherever possible.

biochar An organic charcoal formed from burning biomass in reduced oxygen.

biodiesel A substitute for diesel fuel that is produced from vegetable oils or tallow.

bioelectricity Electricity derived from *biomass*.

bioenergy Energy derived from *biomass*.

biofuel Fuel derived from *biomass*. Some usages limit it to liquid fuel from biomass.

biomass Recent organic material, either plant or animal. Its stored solar energy may be converted into useful energy either by direct combustion or by first converting it into more useful forms by such processes as gasification, fermentation/distillation, anaerobic digestion and *pyrolysis*.

biosequestration The capture of CO_2 by plants by means of *photosynthesis* and the resulting storage of CO_2, for different periods of time, in the plants themselves, leaf litter and soil.

border tax adjustment Payment by the government of a rebate to an export industry to *offset* the increase in production costs caused by a tax or emissions trading or other scheme. Also, applying a levy to emissions-intensive imports to *offset* any significant carbon price disadvantage faced by competing local producers.

business-as-usual (BAU) scenario An energy use scenario that starts from the present pattern of energy use and places no environmental constraints on future economic activity or technology choice.

cap-and-trade An *emissions trading scheme* that places firm limits or 'caps' on total emissions in future years.

carbon dioxide capture and storage/sequestration (CCS) The capture of CO_2 from large point sources of emission and its compression and injection into storages, such as underground geological formations, or oceans, or vegetation, or industrial processes. (Compare *Geosequestration.*)

carbon emissions A shorthand for *CO_2-equivalent* greenhouse gas emissions. Includes all human-induced greenhouse gases, including those that do not contain carbon.

carbon leakage Occurs when a greenhouse gas emitting industry moves from a developed country, as a consequence of government policies in that country, to a less developed country, where the industry allegedly increases its emissions.

carbon tax A tax on a fossil fuel that is proportional to the fuel's carbon content.

champion A high-profile person who is identified with a cause, for example Al Gore, James Hansen. The champion may or may not also be an *organiser.*

Clean Development Mechanism (CDM) A *flexibility mechanism* under the Kyoto Protocol in which industrialised parties to the protocol implement projects that reduce or absorb emissions in non-industrialised countries, in return for certified emission reductions.

climate action group A type of *climate movement organisation* that is devoted wholly to reducing greenhouse gas emissions.

climate action movement A collection of *non-government organisations* and individuals working to reduce greenhouse gas emissions.

climate movement organisation A generic term for a *non-government organisation* wholly or partly devoted to reducing greenhouse gas emissions substantially.

CO_2-equivalent emissions A means of comparing the global warming potentials of all greenhouse gases.

coal seam methane Methane gas found in coal-mines. Chemically the same as natural gas.

cogeneration See *Combined heat and power.*

combined cycle A type of power station with two stages. Waste heat from the first stage, a gas turbine, is used to produce steam for the second stage, a boiler driving a steam turbine. Fuel can be gas, coal or *biomass*, however in practice is usually gas.

combined heat and power A power station in which the waste heat is utilised either for heating water, space heating or industrial process heat.

competitive market Textbook definition is a market in which none of the buyers or sellers can influence prices. But, in terms of comparing with the real world, it might be better to define competitive market as an idealised market, which does not suffer *market failure.*

complementary measures Policy instruments other than a carbon price for reducing emissions. These include *renewable portfolio standards, feed-in tariffs* and tax concessions.

concentrated solar thermal power See *solar thermal electricity.*

Contraction and Convergence A process in which an international greenhouse target, below the existing level of emissions, is achieved by developed countries reducing their per capita emissions and developing countries increasing their per capita emissions, until every country has the same average per capita emissions.

cost curve A graph that ranks *abatement* options according to cost, while showing the quantity of abatement that could be achieved by each option at its own cost level.

distribution line Power line for local distribution of electricity (for example, in suburbs) at low voltages.

economically efficient Less expensive for the same outcome.

efficient energy use (sometimes shortened to 'energy efficiency') Using less energy to provide the same amount of *energy services*: for example, by insulating one's home, or using fluorescent light instead of incandescent, or replacing a fuel wasting car with a fuel efficient car.

El Niño A natural climate phenomenon in which the eastern tropical Pacific Ocean and hence the rest of the world is warmer than average.

Electricity Sector Adjustment Scheme A scheme proposed in the Australian government's White Paper to 'compensate' coal-fired electricity generators for losses in asset values resulting from its *emissions trading scheme.*

emissions-intensive trade-exposed (EITE) industry Industries with high greenhouse gas emissions per unit of production that must compete in international markets.

emissions trading scheme Scheme in which tradable permits to emit pollutants are allocated or auctioned to emitters.

energy Capacity to do work, measured in joules.

energy conservation Reducing the number of *energy services*: for example, heating a home less frequently, or using a car less often.

energy efficiency See *Efficient energy use.*

energy service A task or service that involves energy as an input: for example, home heating, office lighting, transportation. The focus is on the service rather than the quantity and type of energy supplied. To implement the energy service may include *energy efficiency* as well as energy supply.

external cost Something which affects a buyer's or seller's utility or profit which is not included in the prices of goods and services exchanged in the market of interest: for example, the environmental and health costs of burning coal.

fast breeder reactor A type of *fast (neutron) reactor* that produces more *fissile* material (usually as plutonium) than it consumes. It has, around the core, non-*fissile* U-238, which is partially converted into Pu-239 by exposure to fast neutrons.

fast (neutron) reactor A type of nuclear reactor in which the fission chain reaction is sustained by fast neutrons. Such a reactor needs no neutron moderator, but must use fuel that is relatively rich in *fissile* material compared with an ordinary 'burner' reactor, that is, plutonium or relatively highly enriched uranium. Some, but not all, fast neutron reactors are *fast breeder reactors.*

feedback or amplification process When the output of a process increases (positive feedback) or decreases (negative feedback) the input: for example, if global warming melts glaciers to the extent that the Earth reflects less sunlight back into space, then global warming will be increased due to positive feedback.

feed-in tariff Premium long-term electricity tariff paid by a utility for electricity that it must purchase, that is fed back into the grid from a renewable energy source. The price is guaranteed by the government and paid for by electricity consumers, while the quantity of electricity sold is determined by the market. (See also *Gross* and *Net feed-in tariff*.)

fissile (adjective) Element whose atomic nucleus is capable of undergoing nuclear fission (splitting), as the result of being struck by a neutron, is 'fissile'.

fission (nuclear) The splitting of a heavy atomic nucleus into two smaller nuclei with the accompanying release of energy.

flexibility mechanism Under the Kyoto Protocol, a country can *offset* its emissions internationally via three mechanisms: *International Emissions Trading, Joint Implementation* and the *Clean Development Mechanism*.

fossil fuel Coal, oil or gas.

frame (verb) To present an issue to the public in the context and language that you choose.

fugitive emissions Emissions from *fossil fuels* other than those resulting directly from combustion. They include methane gas emitted from coal-mines and leaky gas pipelines, and CO_2 and methane vented from gas fields.

gas In this book, 'gas' denotes either natural gas or *coal seam methane*. It does *not* denote gasoline (petrol).

geosequestration A particular case of *CO_2 capture and storage/sequestration* (CCS) involving storage of CO_2 in underground geological formations, such as depleted oil and gas wells, salty aquifers, and deep unminable coal seams, at depths of at least 800 metres.

grandfathering Allocating emission permits free of charge to emitting industries in proportion to their current or recent emissions.

greenhouse-intensive Adjective describing a task or service that is high in greenhouse gas emissions per unit of output.

Greenhouse Mafia A term which, according to Guy Pearse (see 'Key readings and websites'), the leading greenhouse gas emitting industries in Australia applied to themselves, in the context of these industries' influence upon the government.

Green Power A scheme in which electricity retailers provide customers with electricity from renewable sources for an additional charge.

grid A network of *transmission lines* joining a number of power stations to the main sites of electricity use.

gross feed-in tariff A *feed-in tariff* for which every unit of renewable electricity sold to the *grid* receives the premium tariff.

high-grade uranium ore Ore containing at least 0.1 per cent uranium oxide, U_3O_8.

hydrate Compound involving water molecules.

integral fast reactor A type of *fast (neutron) reactor* that has an onsite *reprocessing* plant.

integrated gasification combined cycle A combined cycle power station that uses gasified coal as a fuel.

intermediate-load power stations Power stations that supply the demand above base-load but below the sharp peaks in demand.

International Emissions Trading An *emissions trading scheme* spanning more than one country that is also a *flexibility mechanism* under the Kyoto Protocol.

isotopes Chemically identical elements that differ slightly in atomic weight and other physical properties (for example, whether they are *fissile*).

Joint Implementation A *flexibility mechanism* under the Kyoto Protocol in which industrialised parties reduce their emissions by means of projects in other industrialised parties.

La Niña A natural climate phenomenon in which the eastern tropical Pacific Ocean and hence the rest of the world is cooler than average.

local centre Part of a city with higher than average residential and job density, having a radius of about 1 kilometre and containing up to about 10 000 people and jobs. A smaller version is sometimes called a 'cluster', or 'compact development', or 'urban village'.

low-grade uranium ore Ore containing 0.01 per cent or less uranium oxide, U_3O_8.

macroeconomic Referring to an economy as a whole, or its major components, as opposed to individual industries, firms, or households.

Mandatory Renewable Energy Target The renewable energy target set up in Australia under the Howard Government. A type of *renewable portfolio standard*.

market failure A situation in which conditions of perfect competition do not apply in a market, with the result that unachieved potential gains are to be made. This can arise because some buyers or sellers can influence prices, or there are externalities, or 'public goods' play a significant role, or there is insufficient information, or institutional barriers to market operation.

mitigation Abatement. Both are technical terms for reducing greenhouse gas emissions.

net feed-in tariff A *feed-in tariff* in which the premium tariff is only paid on the difference between renewable electricity sold to the *grid* and electricity purchased from the grid, provided that difference is positive.

non-government organisation (NGO) A community-based organisation. The definition varies, however, as used here, it excludes big businesses as well as government agencies and includes small businesses and trade unions.

offset Credited reduction in emissions that is purchased from another party to allow the purchaser to increase its own emissions.

organisation A group with some degree of structure. Also, sometimes used in the special sense of the act of organising by an *organiser*.

organiser In the context of social movements, a person who facilitates community empowerment. (S)he may do this by guiding the formation and growth of one or more *climate movement organisations*; helping a group to develop a shared vision, *strategy* and *tactics*; fostering a democratic group structure and decision-making processes; and organising public meetings, workshops, study groups and actions. There are elements of the trade union organiser or cadre in this concept.

peak-load Daily peaks in electricity demand. Also describes type of power station that is used specifically for supplying peaks in demand: gas turbines or hydroelectricity.

peak oil The peak in annual production of oil. Can be applied to the whole planet, a nation or an oil-field.

permafrost Frozen ground. Vast areas exist in Siberia, northern Canada and Alaska.

photosynthesis The natural process by which plants capture CO_2 from the atmosphere and solar energy from sunlight to form carbohydrates, which store the solar energy.

photovoltaic cell A material that directly produces electricity when exposed to sunlight.

primary energy Energy sources obtained directly from the environment, for example, coal, oil, gas, wood, hydroelectricity, wind, solar. Inappropriately called 'energy production' by economists.

pyrolysis Heating a fuel in a limited supply of oxygen, so that it smoulders.

rated power (of power station) Maximum or peak power output recommended by manufacturers for normal operation.

reactor-grade plutonium Plutonium extracted by *reprocessing* from the *spent fuel* of civil nuclear reactors.

rebound effect A situation in which economic savings from *energy efficiency* are invested in using more energy.

renewable energy target A target for renewable energy generation, which may or may not be part of a *renewable portfolio standard*.

renewable portfolio standard A support mechanism for renewable energy that mandates that electricity retailers purchase a minimum proportion of their electricity from renewable energy sources.

reprocessing Chemical process to extract plutonium and unused uranium from the highly radioactive *spent fuel* of a nuclear reactor. Uses remote handling techniques.

solar thermal electricity Electricity generated by using the heat from focused sunlight to boil water to produce steam to generate electricity. Also called 'concentrated solar thermal power'.

spent fuel The highly radioactive used fuel that is removed from a nuclear reactor after a period of operation. It contains fission products (such as strontium-90 and cesium-137), some unused uranium, and *transuranic elements* (such as plutonium-239) created in the nuclear reactions.

stationary energy All energy production and consumption except for transport.

stranded asset An item of economic value, owned by an individual or corporation, that is worth less on the market than it is on a balance sheet due to the fact that it has become obsolete in advance of complete depreciation. For example, a conventional coal-fired power station after a large *carbon tax* has been introduced.

strategy The planning and conduct of long-term campaigns to achieve broad goals.

sustainable development Best known (Brundtland) definition is: to meet the needs of the present, without compromising the ability of future generations to meet their own needs. My own definition is: types of economic and social development that protect and enhance the environment and social equity.

SWOT analysis A strategic method for analysing a proposal in terms of Strengths, Weaknesses, Opportunities and Threats.

tactics Individual steps or tools used in carrying out a *strategy*.

thermal efficiency In the process of energy conversion, useful energy output divided by energy input, usually expressed as a percentage.

transit city A city in which the majority of passenger trips are by public transport.

transmission line Power-line for carrying large quantities of electricity over long distances at high voltage.

transuranic element An element heavier than uranium.

trigeneration Three forms of energy generation from a single unit: electricity, heating (space or water) and cooling.

upstream At points of production or import (of *fossil fuels*).

watt Basic unit of power in SI units, the rate of change of energy generation or energy use over time. 1 watt = 1 joule/sec.

weapons-grade plutonium Plutonium extracted from the spent fuel of a reactor that is operated in a manner to reduce the 'contamination' of Pu-239 with other *isotopes* (such as Pu-238) that would make it less 'efficient' as a nuclear explosive.

KEY READINGS
AND WEBSITES

Unless otherwise indicated, all websites listed here were accessed in March 2009.

Greenhouse science and impacts

Intergovernmental Panel on Climate Change (IPCC) <http://www.ipcc.ch>.

____ (2007c) *Climate Change 2007: Synthesis report. Contribution of Working Groups I, II and III to the Fourth Assessment Report of the Intergovernmental Panel on Climate Change*, IPCC, Geneva, Switzerland, <http://www.ipcc.ch/ipccreports/ar4-syr.htm>.

Pittock, AB (2009) *Climate Change: The science, impacts and solutions*, CSIRO Publishing and Earthscan, Collingwood.

Real Climate <http://www.realclimate.org> (a website by climate scientists).

Schneider, S <http://stephenschneider.stanford.edu>.

____ <http://stephenschneider.stanford.edu/Climate/Climate_Science/CliSciFrameset.html>.

United Nations Framework Convention on Climate Change <http://unfccc.int>.

Greenhouse mitigation: scenarios, economics and policies

Allen Consulting Group (2006) *Deep Cuts in Greenhouse Gas Emissions*, Report to the Australian Business Roundtable on Climate Change, March, <http://www.businessroundtable.com.au>.

Australian Government (2008) *Carbon Pollution Reduction Scheme: Australia's low pollution future*, White Paper, Canberra, <http://www.climatechange.gov.au>.

Centre for Energy and Environmental Markets (CEEM), University of New South Wales, <http://www.ceem.unsw.edu.au>.

Diesendorf, M (2007a) *Greenhouse Solutions with Sustainable Energy*, UNSW Press, Sydney.

Diesendorf, M (2007b) *Paths to a Low-Carbon Future: Reducing Australia's greenhouse gas emissions by 30 per cent by 2020*, Greenpeace Australia Pacific, September.

European Commission, *EU Policies*, <http://ec.europa.eu/policies/index_en.htm>.

European Wind Energy Association (2003) *Wind Energy: The facts*, vols 1–5, <http://www.ewea.org>.

Garnaut, R (2008) *The Garnaut Climate Change Review: Final report*, Cambridge University Press, Cambridge, <http://www.garnautreview.org.au/pdf/Garnaut_

prelims.pdf>.

Laird, P, Newman, P, Bachels, M & Kenworthy, J (2001) *Back on Track: Rethinking transport policy in Australia and New Zealand*, UNSW Press, Sydney.

McKinsey & Company (2009) *Pathways to a Low-Carbon Economy*, Version 2 of the global greenhouse gas abatement cost curve, <http://www.mckinsey.com/clientservice/ccsi/pathways_low_carbon_economy.asp>.

—— (2008) *An Australian Cost Curve for Greenhouse Gas Reduction*, <http://www.greenfleet.com.au/News/An_Australian_Cost_Curve_for_Greenhouse_Gas_Reduction/indexdl_215.aspx>.

Newman, P (2008) *Cities as Sustainable Ecosystems*, Island Press, Washington DC.

Newman, P & Kenworthy, J (1999) *Sustainability and Cities: Overcoming automobile dependence*, Island Press, Washington DC.

Saddler, H, Diesendorf, M, Denniss, R (2007) Clean energy scenarios for Australia, *Energy Policy* 35(2): 1245–56.

Stern, N (2006) *Stern Review: The economics of climate change*, October, <http://www.occ.gov.uk/activities/stern.htm>.

Teske, S & Vincent, J (2008) *Energy [R]evolution: A sustainable Australia energy outlook*, Greenpeace Australia Pacific and European Renewable Energy Council, <http://www.energyblueprint.info>.

UK Department of Trade & Industry (2007) *Meeting the Energy Challenge: A White Paper on energy*, <http://www.berr.gov.uk/whatwedo/energy/whitepaper/page39534.html> to be transferred to Department of Energy & Climate Change <http://www.decc.gov.uk>.

Victoria Transport Policy Institute, British Columbia <http://www.vtpi.org>.

Technologies

Boyle, G (ed) (2004) *Renewable Energy: Power for a sustainable future*, 2nd ed, Open University and Oxford University Press, Oxford.

Sørensen, B (2005) *Renewable Energy: Its physics, engineering, environmental impacts, economics and planning*, 3rd ed, Academic Press, San Diego.

Climate action networks and groups
See table 5.3.

Strategies for social change movements

Alinsky, SD (1971) *Rules for Radicals: A practical primer for realistic radicals*, Random House, New York.

Bobo, KA, Kendall, J & Max, S (2001) *Organising for Social Change: Midwest Academy manual for activists*, 3rd ed, Seven Locks Press, Washington.

Coover, V, Deacon, E, Esser, C & Moore, C (1978) *Resource Manual for a Living Revolution*, 2nd ed, New Society Press, Philadelphia PA.

Ekins, P (1992) *A New World Order: Grassroots movements for global change*, Routledge, London.

Gandhi, MK (1927, 1929) *An Autobiography or The Story of my Experiments with Truth*, Navajivan Publishing House, Ahmedabad.

Green, G (2008) *From Poverty to Power: How active citizens and effective states can change the world*, Oxfam International, Oxford.

Lakey, G (1973) *Strategy for a Living Revolution*, WH Freeman, Boston.

Merton, T (1964) *Gandhi on Non-Violence*, New Directions, New York.

Midwest Academy <http://www.midwestacademy.com>.

Moyer, B (1987) *The Movement Action Plan*, Social Movement Empowerment Project, San Francisco, <http://en.wikipedia.org/wiki/Movement_Action_Plan>.

Moyer, B, McAllister, J, Finley, ML & Soifer, S (2001) *Doing Democracy: The MAP model for organising social movements*, New Society Publishers, Gabriola BC, Canada.

Nonviolence.org <http://www.nonviolence.org>.

Right Livelihood Awards <http://www.rightlivelihood.org>.

Rose, C (2005) *How to Win Campaigns: 100 steps to success*, Earthscan, London.

Sharp, G (1973) *The Politics of Nonviolent Action*, Porter Sargent Publisher, Boston MA.

The Change Agency <http://thechangeagency.org>.

Vested interests

Beder, S (2000) *Global Spin: The corporate assault on environmentalism*, revised ed, Scribe, Melbourne.

Hamilton, C (2007) *Scorcher: The dirty politics of climate change*, Black Inc Agenda, Melbourne.

Pearse, G (2007) *High and Dry: John Howard, climate change and the selling of Australia's future*, Viking, Melbourne.

INDEX

Index

206

Midwest Academy xi, 19, 136–39, 147, 165, 167

Minerals Council of Australia 49

Mondragon 34

monetary system 122

motivating 18, 131, 135, **152–4**, 202

Movement Action Plan (MAP) 19, **154–58**, 165, 170, 192

Movement for a New Society 154

Moyer, Bill 6, 154–55

Murray-Darling 38, 104, 132

naming and shaming 186–87

National Mining Association 23

National Wildlife Federation 142

Natural Resources Defense Council 142

Nature Conservancy 142

Nature Conservation Council of New South Wales 163

network 19, 34, 114, 127, 138–44, 152, 159, 161–65, 176, 178, 185, 198, 200–5

Nike 141

non-government organisation (NGO) 7, 113, 116, 122, 142–43, 156–62, 168, 174, 101, 104, 203–7

non-violence 127, 183–84, see also nonviolent action

nonviolent action 130–31, 155, 176, **183–9**, 193

Nott, Matthew 178

Nuclear Regulatory Commission 42

nuclear power 4, 29, 30–31, 33, 40–43, 46, 55, 61, 66, 79–81, 87, 176, 183

nuclear war 3, see also nuclear weapons

nuclear weapons 41–43, 196, see also nuclear war

Obama, Barack 93, 100, 157–58, 197

offset 81, **104–6**

Olkiluoto 42

OOA 176

OPEC oil embargo 44

Origin Energy 141

Oxfam International 159, 161

organiser 129, 132, **147–52**, 167, 185, 194

peak gas 74

peak oil 74–75, 81, 83

peak-load 45–47, 66, 69–70

Pearse, Guy 25

permafrost 14–15

Pew Center 142, 157, 159, 161

photosynthesis 10, 67, 117

photovoltaic (PV) 26–27, 46–47, 65, **70–71**, 75, 79, 81–84, 95, 111, 115, 117, 179, 188, 191, 193

phytolith 78, 117

plutonium 40–42

policies, key government **90–92**

population

    as driver of emissions 17–18, 21–22, 119

    growth and stabilisation of 5, 57–58, 63

policy 35, 87, 91, **119**

power, political, ix, 2, 6, 8, 17–18, 22, 30, 34, 98, 121, 127, **128–31**, 133, 137–43, 163–65, 172–3, 183, 186, 191–2, 196–97, 204

power-holder x, 5–7, 18, 61, 120–1, 139–43, **152–57**, 164–67, 172–75, 180, 184–86, 191, 198

rail viii, 9, 53, 59, 76, 82, 91, 114, 133

reactor 4, 40–43,

Real Climate 31

rebound effect 44

REDD 116

Regional Greenhouse Gas Initiative 100, 158

regulation 35, 44, 53, 56, 64, 89–90, **112–13**, 125, 133, 171, 207

renewable energy, community based 191–92

Renewable Energy Fund 25, 115

Renewable Energy Target, 26–27, 52, 56, 65, 94–95, 108–9, 142, 163, see also Mandatory Renewable Energy Target

renewable portfolio standard 108–10, see also Renewable Energy Target

Repower America 95

reprocessing 30, 40–41

research 5, 24, 26, 53–56, 67, 75–76, 78, 91, 98, 107, 115, 118, 120, 140, 150, 155, 160, 163, 175, 191, 206

Right Livelihood Award ix

Rockwood Leadership Program 167

Roosevelt, Franklin D 167

Rose, Chris 169–70,

Royal Society 31

Ruckus Society 167

Rudd, Kevin 24–25, 28, 55, 109, 176–77, 194

RWE 29

scenario 44, 50, 78–85, 92–94, 138–39

sceptic, see denier

Schneider, Stephen 31

Schwarzenegger, Arnold 205

Sellafield 30, 40–41

shareholder action 185–87

Sharp, Gene 127, 130–31, 189

Shell 23, 134

Shiva, Vandana ix

Sierra Club 161, 189, 204–7

Sizewell B 41

social defence 148

social justice ix, 7–8, 16, 113, 131, 152, 159, 184, 205, see also just transition and ethics

social movement 6–7, 18–19, 85, 131, 138, 140, 143, 148, 154–56, 196, see also climate action movement

solar photovoltaic, see photovoltaic

solar thermal 26, 46, 54, 68–69,

www.ingramcontent.com/pod-product-compliance
Lightning Source LLC
Chambersburg PA
CBHW050705280326
41926CB00088B/2601